# ASTAXANTHIN
# Seafood's Ultimate Supernutrient

By
William Sears, M.D.

Illustrations by Debbie Maze

© 2013 by William Sears, M.D. All rights reserved.

ISBN #: 978-0-9792353-2-0

# TABLE OF CONTENTS

# Forward

Though I've spent my medical career with an open mind to new interventions and treatments in the field of health and medicine, the scientist side of me has always been a skeptic - show me the research.

With a dizzying array of new products, claims and information in the natural products and dietary supplement industry, it's often difficult to discern the wheat from the chaff. Sure, a strong marketing campaign or successful viral video could lead to a brief bump in awareness and sales, but the acceptance and long-term sustainability of an ingredient or product is really dependent upon strong, ongoing scientific support.

When I first heard about Hawaiian Astaxanthin, I was surprised by the amount of scientific research that already existed for this impressive ingredient. What's been even more encouraging is the rate at which new discoveries, mechanisms of action, and uses are uncovered. I was also surprised by the potency of Astaxanthin, which only requires 4 - 12 milligram dose ranges. This is in sharp contrast to many natural ingredients, especially in the herbal world, that require several grams to achieve positive research outcomes.

The effect, and the favorable dosing, opens up a whole world of opportunities for Hawaiian Astaxanthin. Not only can it be used alone, but is a valuable addition as a combination ingredient for so many areas of health, such as brain, joint, eye, heart, immune support, and even cosmetic applications, among others.

Besides all the promising data in the research and the favorable dosing amount, the purity, availability, and sustainability of Hawaiian Astaxanthin are real bright spots when compared with other ingredients. The growing and cultivation of Hawaiian Astaxanthin in a pristine climate, with pure water and raw materials help create a finished product which is both potent, reliable, and essentially devoid of pesticide residues and heavy metals that can be problematic in so many other ingredients in the industry. It's a relief to know that we all have access to a pure and reliable ingredient.

The future is indeed bright for Astaxanthin. It's going to be exciting to see its continued development and progress going forward.

Jason Theodosakis, MD, MS, MPH, FACPM, Author of the *New York Times #1* Bestselling Book, *"The Arthritis Cure"* and Clinical Associate Professor of Family and Community Medicine at the University of Arizona.

Disclaimer: The University of Arizona as a matter of policy does not endorse specific products or services. Dr. Theodosakis' credentials as a Professor are for identification purposes only. These statements have not been evaluated by the Food & Drug Administration; they do not represent a diagnosis nor a recommendation for treatment, cure or prevention of any disease.

## A NOTE TO MY READERS

We all search for the secrets to health. Reading this book will enlighten you about one of nature's top secrets: People who eat the most antioxidants (natural anti-wear-and-tear nutrients) tend to enjoy the best health. You are about to embark on a journey to the land and sea of antioxidants and learn which type is the most healthful.

Among my professional colleagues I'm known as the show-me-the-science doctor. I won't take, or eat, or prescribe any medicine or start a health habit that is not backed up by science. My family, my patients, and my readers trust that science-based sincerity. I'm also known as the science-made-simple doctor. In fact, when I'm paired with researchers in talks at medical meetings we jokingly call this the scientist and simpleton talk. So, readers, I assure you this book is not only science-based, but the science is simple and fun to read. Enjoy!

## DR. SEARS' LEADING TITLES INCLUDE:

### SEARS PARENTING LIBRARY

*The A. D. D. Book*
*The Attachment Parenting Book*
*The Autism Book*
*The Baby Book*
*The Baby Sleep Book*
*The Birth Book*
*The Breastfeeding Book*
*The Discipline Book*
*The Family Nutrition Book*
*The Fussy Baby Book*
*The Healthiest Kid in the Neighborhood*
*The Healthy Pregnancy Book*
*The N. D. D. Book*
*The Portable Pediatrician*
*The Pregnancy Book*
*The Premature Baby Book*
*The Successful Child*
*The Vaccine Book*

### SEARS CHILDREN'S LIBRARY

*Baby on the Way*
*Eat Healthy, Feel Great*
*What Baby Needs*
*You Can Go to the Potty*

### OTHER SEARS BOOKS

*Prime-Time Health*
*The Omega-3 Effect*

## CHAPTER 1

# THE ASTAXANTHIN EFFECT IN NATURE

## MY SECRET FOR FINDING THE SECRET TO HEALTH

Following a life-threatening illness at the age of 57, I searched for the top secrets to health – and found them. At the prime of my life with eight children and then five grandchildren, in addition to a thriving medical practice and publishing career, I had a lot to live for.

Fifteen years later, at the age of 73, I'm enjoying wonderful health with the blood pressure and blood chemistry of a youngster, and I take no regular prescription medicines. I enjoy the simple health goal that everyone wants: Everything works and nothing hurts. Naturally, I was motivated to answer the question: What do people do who live the healthiest and longest? The answer: Go fish! In searching for the top health tips, I found the top health food: *seafood.* The incidence of just about every illness goes down with the more seafood you eat. People who eat more food from the sea enjoy these head-to-toe health benefits:

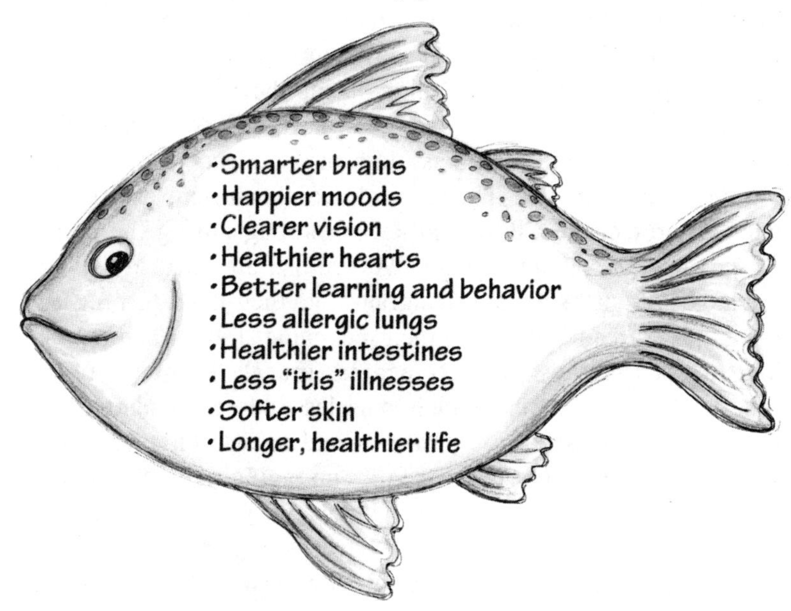

**What nutrients in seafood make it so special?** In my search through hundreds of scientific articles, the health and longevity nutrient that kept getting scientists' top vote was the *omega-3 fat* found in fish oil. This discovery led to my book: *The Omega-3 Effect* (Little, Brown, 2012). A six-ounce fillet of wild Alaskan sockeye salmon, in my opinion, wins the top health food award because it packs the most healthy nutrients per calorie. While the omega-3s, vitamin D, protein, and calcium in seafood all deservedly get the most press, there is an overlooked, undervalued, and least understood nutrient in pink seafood that is one of nature's most powerful health foods – Astaxanthin.

## WHY SALMON ARE PINK

As I was fishing for knowledge on why seafood is the perfect health food, I had the opportunity to go fishing with a real Alaskan fisherman, Randy Hartnell, owner of Vital Choice Seafood Company. While watching wild salmon swim mightily upstream during their marathon race, I was amazed at their strength and stamina to keep going without tiring. One morning while watching this salmon race, I asked Randy, "Why are salmon pink?" His answer led to the writing of this book.

Wild salmon feed on krill, small fish, and algae that are plentiful sources of the pink pigment Astaxanthin. When that internal switch clicks on and drives these fish to leave the ocean and return to their birth river to spawn, as if guided by some internal GPS, they stop feeding during their vigorous journey upstream and live on their rich stores of fat and other nutrients accumulated for their race. Before their final sprint, the salmons' flesh turns a deeper pink because it stores Astaxanthin from their diet. This powerful pink nutrient acts like an internal bodyguard to protect their flesh and keep it strong during the marathon. The harder they work, the pinker they get. As the fish digest their own stored fat for energy, a large dose of Astaxanthin migrates into their flesh, turning it pinkish-red and giving the upstream racing fish their nickname "red salmon."

**Dr. Mother Nature protects against muscle fatigue.** To give salmon the strength to swim upstream, their muscles undergo oxidative stress, which simply means lots of tissue wear and tear and muscle fatigue. Think about it, these mighty fish continue swimming miles up raging rivers for up to seven days. What makes these heroic swimmers accomplish the greatest athletic feat in nature without damaging their muscles? The answer is: Natural Astaxanthin. Without this pink tissue protector, it's unlikely the salmon would reach their destination.

*Salmon swimming upstream.*

It's interesting that Natural Astaxanthin (an antioxidant) is found in the highest concentrations in the muscles that are stressed the hardest (that produce oxidants), such as the flesh of wild salmon. Isn't Dr. Mother Nature clever to instill the strongest antioxidant in the muscles that endure the highest oxidative stress? Oxidative stress is the biochemical quirk that makes us sick, tired, and old.

Deep down I was concluding: If Astaxanthin is fabulous for fish, could it also be healthful for humans? The scientist in me said, I wonder if the same nutritional principle applies to fish as what mom preached about fruit and veggies: *The deeper the color, the better it is for you*. The reddish-pink color in wild salmon comes from the nutrient Astaxanthin, one of nature's most powerful *antioxidants*. Antioxidants are nutrients that neutralize *oxidants*, wear-and-tear biochemicals the body produces not only during intense exercise, like the salmon run, but also as byproducts of the daily energy production during tissue growth and repair. (More about antioxidants on Page 14.) This pink "medicine" is many times more potent than the antioxidants vitamin E and C. So, mom's wisdom of put more color on your plate pertained not only to fruits and vegetables, but also to fish.

**Remember, *Astaxanthin* = Antioxidant.**

After fishing for, and catching, the best nutritional information about the best fish from the most nutrient-filled waters of the world, Alaska, I wanted a second-helping of why this salmon-saving nutrient was so healthy for humans.

## FISHING IN HAWAII

Next, I went from pink fish to red ponds. In fishing through medical journals, I kept seeing the term "Hawaiian Astaxanthin." So, off to the Big Island of Hawaii to the world's largest Astaxanthin farm. My first impression as I gazed upon the twenty-five blood-red ponds was what a field of dreams this is to learn what Astaxanthin is and why it's so healthful. The eye-opening scene made me wonder: If fish eat this supernutrient from Dr. Mother Nature to stay healthy, and we eat the fish that eat the supernutrient, then it must be good for us. Three more visits to these red ponds confirmed what scientists say: Astaxanthin is one of the world's best-kept health secrets.

*Aerial view of Astaxanthin farm, Kailua-Kona, Hawaii. Photo courtesy of Cyanotech Corporation.*

"The number one supplement that you've never heard of that you should be taking" – *From the Dr. Oz show.*

**Pinking up pale salmon.** Why are some salmon pinker than others? The nutritional wisdom "we are what we eat" also applies to salmon. One reason wild salmon are pinker than the farmed ones is the wild ones eat real Natural Astaxanthin; the farmed ones eat Synthetic Astaxanthin. Natural Astaxanthin has *health* effects; Synthetic Astaxanthin shows *cosmetic* effects.

Suppose you were in the business of fish farming. Since the main goal of business is profit and not health, you would feed your fish the cheapest food that still makes them grow and look like real fish. Fish are picky eaters. They won't grow or even survive unless they eat at least some seafood. Their genes program them to be that way. These farmed fish make their way to the restaurant menu as "Atlantic salmon." Early on in your fish farming experiment you do a health check. You notice that some of your salmon are pale and not pink. These salmon won't sell because they don't look like real salmon. So you call your local fish doctor who sends you a "pill" to pink up your pale salmon to look healthier. That "pink up" pill is Synthetic Astaxanthin. The problem with this pseudo pink stuff is that it is like other man-made synthetic "food" – it may not be as healthful for the body as what Dr. Mother Nature made. A four-ounce fillet of wild sockeye salmon, which is the best-known salmon in the Pacific, contains 4 milligrams of Astaxanthin. You would have to eat two pounds of farmed Atlantic salmon to get that much Astaxanthin.

Another concern is that Synthetic Astaxanthin is made from petrochemicals. That's right – pretty much the same stuff you put in the crankcase of your car. Not only is Synthetic Astaxanthin from a questionable source, but it's also a different animal in terms of how it works. For example, in one antioxidant experiment at Creighton University, Natural Astaxanthin from algae was twenty times more potent as an antioxidant than Synthetic Astaxanthin (Bagchi, 2001).

**See the difference in seafood.** Next time you're in a supermarket that has a wide variety of seafood, compare a fillet of wild Pacific salmon to a farmed Atlantic salmon. The wild one that lived the wild life is pinker, almost red. The factory-made pale pink color (Synthetic Astaxanthin) in the farmed fish doesn't behave in the body as well as the deeper-colored pink pigment (Natural Astaxanthin) in the fish the fisherman caught by hook or net.

**A tale of two fish: The wild one and the farmed one.** Millions of years of recipe-perfecting have produced the perfect food for growing perfect fish. The wild one, by some still mysterious primitive instinct, tells itself: "It's my time to find my birthplace, and I must swim there no matter what. Nothing will hold me back." But like muscle-training for marathons, the wild one knows she won't make it to her destination without storing up nutrients for the vigorous voyage. She must live off her own flesh that gives of itself for the good of getting there. So just before the race, the wild one goes on a feeding frenzy, gobbling up all the pink powerful foods (krill and Astaxanthin-rich sea plants) she can get to protect her muscles from self-cannibalizing during the upstream run. This is when fishermen try to catch the wild salmon: when they're at their pinkest, fattest, healthiest, and tastiest, just before they start to make their run upstream. The depth of their deep color – and their highest nutrient concentration – occurs while they keep themselves in a kind of natural holding tank days or weeks before their "go button" prompts them to run for the fishy finish line.

The farmed one, on the other hand, sits around eating man-concocted recipes of food all day. Even if inclined to swim upstream and join the wild ones in the race, she can't escape from the pen to do so. So, the penned-up less-perfect fish gets fatter from a substance that is less healthy for it than what the wild ones eat, and ultimately less tasty when humans eat it.

Here's another mother-truism that you probably never understood even if you did listen: "You're not only what you eat, you're also what the animal/fish eats." If you eat the wild one, you also eat those wild nutrients the wild one eats, and those nutrients become you. If you eat the farmed one, well, you get the picture. (See more about Natural vs. Synthetic Astaxanthin, Page 51.) Wild fish have a healthier nutrient profile *because they eat real food.* We humans could learn that lesson from our sea friends.

**From colorful fish to colorful humans.** Now that you've followed these fish stories and value the pink pigment that keeps fish healthy, let's learn more about what Astaxanthin is and what healthful effects it has on your body.

What works in nature also works in humans. Could Astaxanthin be as healthful for human bodies as it is for salmon? Yes! Since I'm an exercise addict, I realized that the more my body moves, the more Astaxanthin I need. So, I used this simple logic to conclude I needed to eat more Astaxanthin:

- Dr. Mother Nature proved it's safe and effective for hardworking tissues
- My body is a hardworking tissue
- Science says it's healthful
- Common sense says it's healthful

**Pass the Astaxanthin, please!**

## ALL ABOUT ASTAXANTHIN: WHAT IT IS, WHAT IT DOES

What is Astaxanthin, and what makes it so special that highly exercising tissues can't live well without it? Astaxanthin is a *nutraceutical*, a nutrient that has proven pharmacological health benefits. Astaxanthin is correctly called "The King of Carotenoids." While you may not know what carotenoids are, you've probably eaten a lot of them in the past 24 hours. Carotenoids are the pigments that give many of the healthy foods we eat their color: the lycopene in the red tomato, the zeaxanthin and lutein (you've probably heard about these for eye health) in the yellow corn, and the beta carotene in the orange carrots. While there are over 700 different carotenoids in nature, you've probably only eaten a few of them.

Where in the world of nature is Astaxanthin? Besides salmon gobbling it from the ocean's smorgasbord, it is also what gives pink flamingos their color. Flamingos eat algae that contain the carotenoid zeaxanthin and the orange carotenoid beta carotene. Then their bodies convert these carotenoids into the pinkish-red carotenoid Astaxanthin. Among the plant life of the sea, algae is the most prevalent. The alga that produces Astaxanthin is called Haematococcus pluvialis. Haematococcus algae are one-cell plants; they're like microscopic little red balls that can stay alive for years, surviving intense sunlight and other forces of nature such as the summer sun and winter cold. What makes them so strong is their accumulation of Astaxanthin, also known as *"The Great Protector."* The powerful pigment turns the algae red and, in so doing, not only protects them from damage but enables a ready food supply for hungry fish.

Astaxanthin can be found in sea plants and sea animals. It is most prevalent in algae and phytoplankton, or sea plants. Any sea animal that has a reddish or pinkish color, such as salmon, trout, lobster, shrimp and crab, contains Natural Astaxanthin. These seafood eat krill and other small sea organisms that eat Astaxanthin-containing algae and plankton as a major part of their diets. The sea animal that has the highest concentration of this king carotenoid is salmon. Astaxanthin concentrates in their muscles and makes them the endurance heroes of the animal world. If it weren't for Astaxanthin, not only would the salmon cease to be a delectable delicacy on our tables, but they'd be pale, worn out, and tired all the time. Sounds like some humans who are antioxidant deficient.

Now let's take a trip through the body, even as deep as the cellular level, to learn about what one of nature's most powerful and colorful pigments can do for you.

## SCIENCE SAYS: ASTAXANTHIN IS AWESOME

While studies of the health effects of Astaxanthin are still in their infancy, here's a summary of what science says this pink powerhouse can do for you:

- Protects the *brain* against dementia
- Protects the *eyes*
- Modulates the *immune system*
- Protects *skin* from UV sun damage
- Protects *cells* against DNA damage, the precursor to cancer
- Improves endothelial function
- Reduces insulin resistance
- Lessens atherosclerosis
- Reduces inflammation
- Improves blood lipids: increases HDL, lowers triglycerides
- Reduces tissue wear and tear during intense exercise

Some of this science is in its infancy, meaning many of these studies have a qualifier: "Further research is needed to substantiate these findings."

### WHO NEEDS ASTAXANTHIN?

***If you have any of these health concerns, you need the Astaxanthin effect:***

- Excess weight
- "-itis" illnesses: arthritis, bronchitis, colitis, dermatitis, tendonitis
- Cognitive difficulties
- Cardiovascular disease
- Vision problems
- Fatigue
- Athletic muscle fatigue
- All of the above

## HOW WE GET SICK – HOW WE STAY WELL: MY "STICKY STUFF" EXPLANATION OF ILLNESS

When I was putting together my own health plan, I thoroughly researched the question: What exactly happens in the body that causes it to get sick, hurt, and age. I do my best thinking while swimming. When I draw a blank while thinking, I decide: "I'll swim on that!" For me, the free-flowing movement of the body pushes aside distracting and competing thoughts and allows clearer thoughts to flow. It was on a swim that I came up with my **sticky-stuff explanation of illness**. I wanted a simple, memorable, teachable, yet accurate explanation of how our body parts break down and how we can keep them healthy. Nearly every illness and pain at every age is caused by accumulation of sticky stuff in tissues and cured by preventing or removing the sticky stuff from the tissues.

Common examples of how sticky stuff accumulates in tissues, causing illnesses and shortening longevity are:

- Early aging: wear and tear, wrinkles, gray hair
- Cataracts
- Vision loss
- Cardiovascular disease: hardening of the arteries, atherosclerosis
- "-itis" illnesses, the ABCD's: arthritis, bronchitis, colitis and cognivitis (Alzheimer's), and dermatitis

As a physician I try to explain health and wellness in simple terms. My patients like my "sticky stuff" explanation: When you accumulate too much sticky stuff in your body, you get sick and are destined for the D's: *D*isease, *D*iabetes, *D*isabilities, *D*octors, more *D*rugs, and the final D that occurs earlier than necessary in many people. When you keep excess sticky stuff out of your body, you decrease your chances of getting these D's and are likely to stay healthy – at all ages.

The good news is the natural nutrient in seafood, *Astaxanthin*, can help protect you from all this sticky stuff accumulation. Here's how:

**Shun the sticky stuff.** What actually is this sticky stuff? Medically speaking, this sticky stuff goes by the chemical-sounding names *oxidation* and *inflammation*, or the "shuns." Simply speaking, for good health and longevity, shun the "shuns." Let's start with oxidation. Our bodies are oxygen-burning machines. Every minute countless biochemical reactions throughout the body generate trillions of particles of "exhaust." These sticky particles of exhaust are known as *oxidants*, also called free radicals. Trillions of times a day these oxidants hit our tissues like a steady rainfall rusting away our cells. The rusting caused by these oxidants not only contributes to chronic diseases, such as the "-itis" illnesses, but eventually leads to wear and tear, such as hardening of the arteries, stiff joints, blurry vision, and wrinkled skin.

**Antioxidants are anti–sticky stuff.** Normally, our bodies soak up these oxidants by producing anti-rust chemicals called *antioxidants*. But when the body builds up more oxidants than antioxidants, rust accumulates and increases the wear and tear on our tissues. Unfortunately, as we age our bodies tend to produce fewer antioxidants. So, as we get older, we need to eat more foods that are rich in antioxidants.

## ASTAXANTHIN SLOWS AGING

Once upon a time it was thought that the ailments of aging were mostly due to tissue wear and tear over time. The aging body was viewed like a machine that simply wears out because of all the parts rubbing together: Blood flowing through blood vessels eventually wears out the vessel lining; joints being jarred all day long wear out the joint lining; and skin exposed to the elements, such as excess sunshine, gets wrinkled. This simplistic explanation of aging is only partially true.

An updated and scientific explanation of aging is the tissue damage caused by the "shuns," mainly oxidation and inflammation. This sticky stuff triggers a cascade of tissue damage that causes tissue to get too old too fast. (Another "shun," *glycation*, happens when sticky stuff accumulates from eating too much added sugar.) The Astaxanthin effect slows aging by decreasing the "shuns."

**– The AAA Effect of Aging –**

***When we age, our bodies produce less antioxidants, so we need more Astaxanthin.***

The Astaxanthin effect is particularly healthful for seniors:

- The anti-wrinkle effect (see Pages 48-49)
- The anti–sun damage effect (see Pages 48-49)
- The anti-inflammatory effect (see Page 41)

Does Astaxanthin help you not only live better but live *longer*? Science is not yet sure. Yet, an interesting 2011 study showed some intriguing life-lengthening effects. To test whether or not Natural Astaxanthin could extend longevity, scientists fed the supernutrient to worms from larval to adult stage. Astaxanthin protected the worms from wearing out, extending their lives by 16% to 30% (Yazaki, 2011).

**STICKY STUFF CYCLE OF ILLNESS:**

Sticky stuff accumulates in tissues – causing excess oxidation, which leads to inflammation.

*Inflammation.* Because the tissues are now changed, or damaged, by the accumulation of sticky stuff, the immune system thinks the oxidized tissues are foreign and attacks them, causing another "shun" – inflammation. The inflammation just adds more sticky stuff on top of the oxidation. Our natural immune system (such as protector cells and germ-fighting cells) gets its signals mixed up and wrongly concludes: This sticky stuff doesn't belong here, so we must attack it. And the resulting metabolic mess causes more inflammation.

Inflammation plus oxidation lead to more sticky stuff, which leads to more inflammation, which leads to more sticky stuff, and the cycle continues, causing the D's – disease and disabilities – especially in sensitive tissues, such as the blood vessel lining, joint lining, intestines, and lungs.

### WHERE WE GET OXIDANTS

**The five main sources of sticky stuff oxidants are:**

1. Normal body chemical reactions, or metabolism, which generate oxidants.
2. Excessive ultraviolet ray exposure to skin.
3. Air pollutants we breathe.
4. Excess exertion where we generate more muscle exhaust than the antioxidants can keep up with.
5. The immune system, which attacks one "shun" – oxidation – but produces another "shun" – inflammation.

**Understanding inflammation.** *Inflammation* is the current healthcare buzzword. When your body is in inflammatory balance, you enjoy health and longevity. When your body is out of inflammatory balance, you get sick. So, the secret to good health is keeping your body in inflammatory balance.

To understand inflammation, which literally means "on fire," let's use the road repair analogy. Suppose there was a road well-traveled in your body, say the lining of a blood vessel that has blood (traffic) flowing through it, or even the lining of a joint, such as a knee joint, that's flexed thousands of times a day. There is a lot of rubbing that goes on in these tissues, so that eventually the lining gets rough. Your body has a marvelous built-in maintenance system. It sends out the message that the road is rough and needs repair. Maintenance engineers are deployed. These are actually inflammatory chemicals (repair cells) that go by various biochemical names, such as *cytokines*. Their job is to mobilize the repair crew to fix the wear and tear on the road – fill the potholes and smooth the surface. Yet, sometimes these overzealous maintenance engineers overreact, or over repair. The potholes are overfilled, leaving a bumpy road and other rough surfaces.

So, the body mobilizes even more maintenance engineers, but during their repair they add more build-up of sticky stuff, or bumps, in the road. Eventually, the road gets so bumpy that traffic comes to a halt (e.g., a clot forms in the artery). This leads to dysfunction of the organs whose blood is supplied by the artery (resulting in a stroke or heart attack), or joints get painfully frozen and need repair, such as knee or hip replacement. These common disabilities could have been prevented, or at least delayed, if the body's repair system would have been in balance with the body's wear and tear. That simple, but underappreciated, concept of *inflammation balance* is one of the top secrets to good health. (On Page 41 you will learn how the Astaxanthin effect helps quench the "fire" of inflammation.)

## GET THE CHEMICAL PICTURE

What makes the Astaxanthin molecule so special? The best antioxidants are those that get into the most tissues. As you will learn in more detail on Page 26, some antioxidants get into fat tissue, and others get into blood and muscle. Here's where Astaxanthin shines. It gets into nearly all the body's tissues. Astaxanthin enjoys a chemical perk that sets it above the other carotenoids. It has an extra biochemical formula called a hydroxyl group attached to the ends of the molecule. This smart change of nature enables Astaxanthin to work its healthy way into tissues better than the other antioxidants.

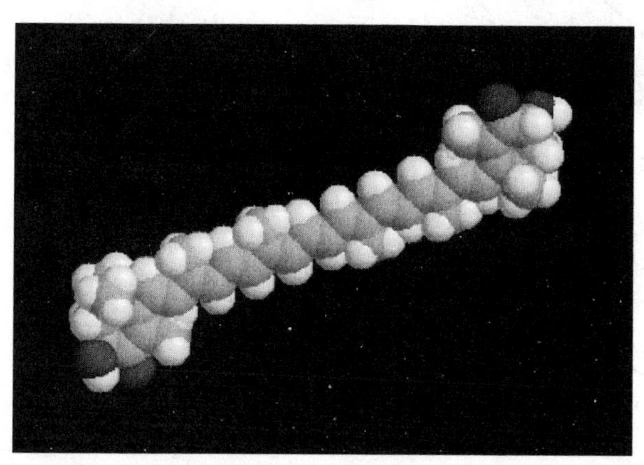

*The Astaxanthin molecule's special hydroxyl group at both ends.*

# CHAPTER 2

## MEET ASTAXANTHIN THE ANTIOXIDANT

I had often heard the terms "oxidation" and "antioxidants," yet like so much biochemical stuff you learn, you forget it until it has personal meaning. Science says: "People who eat the most antioxidants enjoy the best health." When I did a scientific literature search for "antioxidants," Astaxanthin kept popping up, and it was easy to remember: the Astaxanthin effect = the antioxidant effect. So, to fully understand the Astaxanthin effect let's first understand the antioxidant effect.

*Oxidant Quenching*

### ENJOY THE ANTIOXIDANT EFFECT

Every cell in our body, trillions of them, are microscopic engines burning the fuel (food) we eat to generate energy to grow, repair, move, think, and so on. Just as a car engine burning fuel generates exhaust, our body generates exhaust, called *oxidants*, from the fuel – oxygen that the body burns. Accumulation of oxidants, sticky stuff, is what leads to early aging and many illnesses. Oxidation, a natural biochemical process, is our body's way of converting food and oxygen into energy. We can't live without it. But too much oxidation can hurt us. Like excess inflammation, it's excess oxidation that throws the body out of biochemical balance.

Enter *antioxidants*, anti–sticky stuff natural biochemicals, found in the many popular foods your mother made you eat – fruits, vegetables, and seafood. These antioxidants neutralize or suck up the oxidants before the oxidants can do their tissue damage. Think of oxidants as trillions of tiny hot potatoes that need to be passed from one antioxidant to another until they cool off before they "burn" the tissues.

Now that you've learned a bit about oxidants and antioxidants on Pages 9-10, let's go deeper into your metabolism to understand how excess oxidants harm your health and why you need to "anti" these oxidants. Oxidation, like the burning of fuel (such as in a car engine), is like a mini explosion that occurs trillions of times every minute throughout your tissues. When these energy-producing explosions occur, they give off biochemical stuff called *oxidants*. They are also known as *free radicals*, so-called because they behave in the body like a bunch of out-of-control cell damagers. Basically, they are free electrons. In simple yet scientific language, the oxidant effect on our tissues is known as "hits." The hardest-working tissues, such as the brain and eyes, take the most hits. According to University of California antioxidant researcher Dr. Lester Packer, author of *The Antioxidant Miracle*, each cell in our body takes around 10,000 oxidative hits daily. Multiply that by trillions of cells and each day your body takes a lot of "hits."

Antioxidants act like biochemical buddies that pair up with the oxidants to keep them from bullying their way into tissues and causing illness. To help tissues heal, these hits need their buddies – anti-hits, or antioxidants – to keep tissues in antioxidant balance. Otherwise, these oxidants, like mini jackhammers, will keep hitting away at tissues, causing damage. Once the oxidants go haywire and oxidation begins, a chain reaction can occur that generates more oxidants, causing a downward spiral of one health problem leading to another.

Like a well-designed car engine, the body can handle a normal amount of exhaust, or oxidants. Healthfully, a beautiful balance of wear, tear, and repair goes on 24/7 in every part of your body. Yet, our modern lifestyle and diet have caused many of us to have a body out of balance: wear and tear occurs faster than repair. The pollution that goes into our bodies, especially through the chemical air we breathe and the chemical food we eat, shifts the balance toward excess oxidation, or more wear and tear than repair, leading to disease, disability, and earlier death. In a nutshell, we wear out too soon.

Biochemically, sticky stuff, or the oxidants, is the waste product of cellular metabolism. Therapeutically, anti–sticky stuff is known as antioxidants. An unfair quirk of aging is that as we get older, most of us continue to produce just as many oxidants, but our bodies decrease their production of antioxidants. When oxidants equal antioxidants, you remain healthy. When oxidants outweigh antioxidants, you get sick.

**A lesson from daily living.** Back to the car analogy. What happens if you expose metal to the wear and tear of the elements? It oxidizes, or rusts. So, if you put paint (an antioxidant) on the metal, it keeps it "young." While it doesn't sound so sexy, aging is rusting.

Any health plan needs to begin with what you can *do* and *eat* to neutralize the oxidants that are formed every millisecond in your body. Simply put, illness and aging are caused by oxidants, or "-itis" illnesses; therefore, eat more antioxidants. You can do that!

**A lesson from daily eating.** Here's a graphic home exercise to appreciate how antioxidants, such as Astaxanthin, help keep sticky stuff from forming in your tissues. Take either one of the A plants that are health foods: apples or avocados. I prefer to do this demonstration with an avocado to make my point, since another nutritional principle is the more healthy fats a food contains, the quicker it spoils (oxidizes). Halve the avocado. Pour lemon juice (an antioxidant, mainly vitamin C and other biological protectors called bioflavenoids) on one half and leave the other half exposed to the air unprotected. Reexamine your experiment in around six hours. The antioxidant-protected avocado half looks fresh, young, and healthy. The unprotected half looks unhealthy, wrinkled, old, and even rusty. Without antioxidant protection, the avocado got sick and aged.

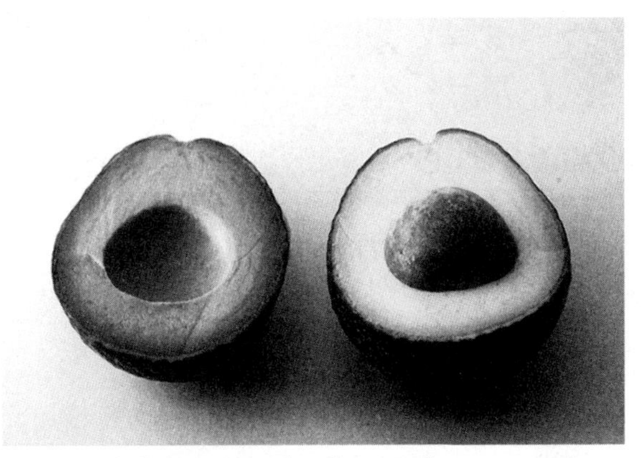

*Avocado halves natural and antioxidant-protected.*

The avocado analogy is what happens to our bodies when we either expose our tissues to oxidation or don't eat enough antioxidants. As the avocado rusts, so do our tissues, such as wrinkled skin, weakened muscle tone, rough and aching joints, blurred vision, and clogged arteries. (You will learn more about this later.) The illustrations on the next page demonstrate the shift in your wellness as you increase or decrease the intake of antioxidants:

## A Top Secret to Health and Longevity: Antioxidant Balance

Shift your balance toward eating more antioxidants and decreasing your exposure to oxidants.

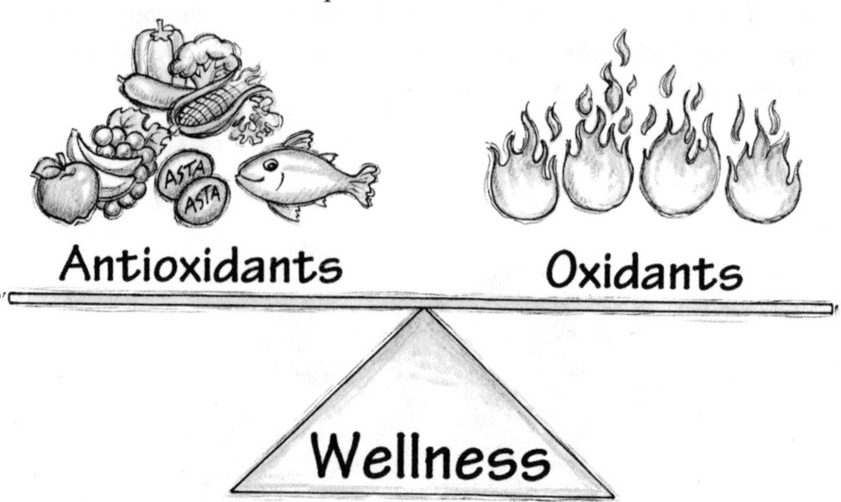


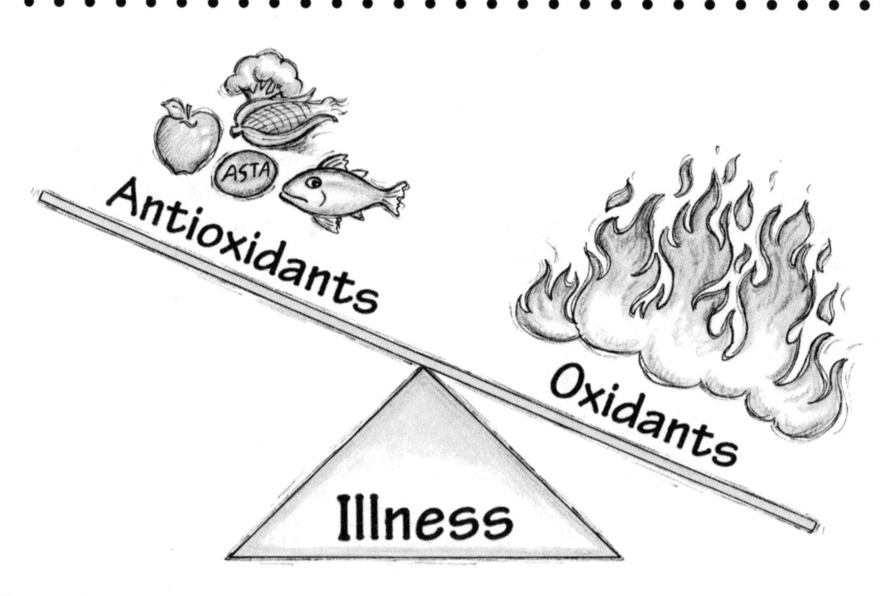

The important thing to consider is that you don't all of a sudden get one of the "shuns." They build up over a long time. Notice the artery below. I use an artery as one of the main tissues affected by the "shuns." In fact, I directed my health plan toward healthy arteries, intuitively knowing that if every organ of the body is only as healthy as the blood vessels supplying it, then if I have healthy blood vessels, I'll have healthy tissues and healthy organs.

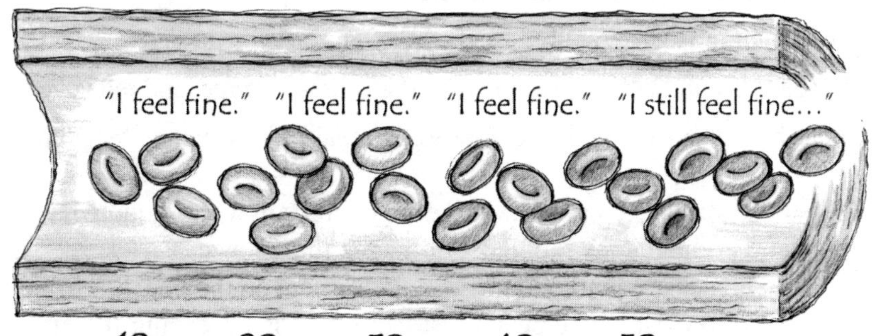

## WHY ASTAXANTHIN IS THE MOST POWERFUL ANTIOXIDANT

Now that you understand what antioxidants are and what they do, let's choose the best ones for you.

**An imaginary antioxidant contest.** Suppose all the top antioxidants got together for a contest. Let's call it a health-lifting competition. Many of the contestants were familiar favorites:

- the A, B, C, D, E team (the vitamins)
- the flavonoids (represented by the berries)
- the carotenoid team (represented by the yellow, orange, and red vegetables)
- and, finally, a relative newcomer to the health food scene (Astaxanthin)

"Who is the strongest antioxidant?" each contestant was asked. All claimed to be the leader, yet the judges shouted, "Show me the science."

How do you measure antioxidants and their effect on the body? The current blood tests are not precise. The best a test can claim is observing the effects of antioxidants in laboratory test tubes or in experimental animals, such as rats and mice. Then scientists make the leap of faith to translate lab and animal studies into what each of these antioxidants does in the human body. When the judges examined the scientific literature, they came to two conclusions: (1) "Astaxanthin beats all of the other teams." (2) "You're all winners!" If Dr. Mother Nature were a real scientist she would conclude: None of you exist by yourself in my laboratory. You all are winners because nowhere in nature do nutrients act alone. They always partner with one another for a synergistic effect. (See the explanation of the nutritional principle of synergy on Pages 51-52.)

**Tests show Astaxanthin is more powerful than other popular antioxidants.** This does not mean you should take Astaxanthin instead of other antioxidants, especially those found in fruits and vegetables, since all have their unique effect. Notice from the graph below that Natural Astaxanthin is over 5,000 times stronger than vitamin C, and much stronger than two other well-known antioxidants, co-enzyme Q10 and green tea catechins.

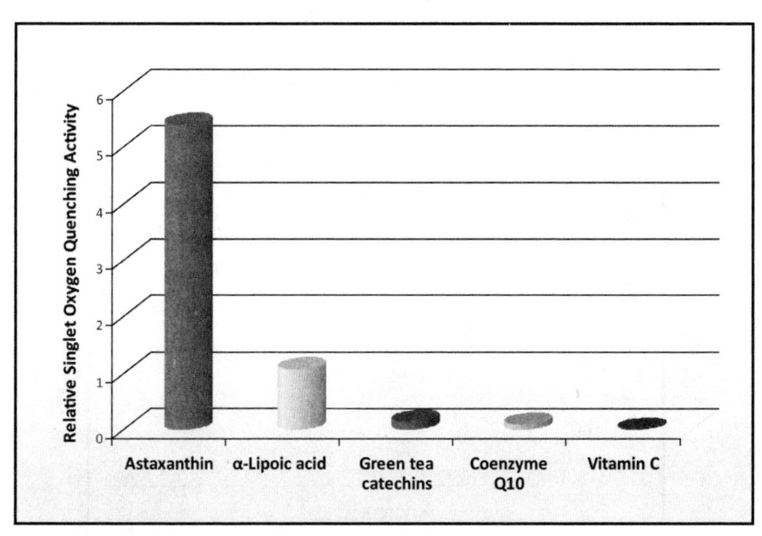


*Y. Nishida, E. Yamashita and W. Miki 2007*

If you're interested in how these antioxidant effects are measured, here's one way it's done. You measure the antioxidant's *singlet oxygen neutralizing capability*, meaning how powerfully it neutralizes one of the most damaging oxidants called singlet oxygen. In biochemical speak, this neutralization is called "quenching," sort of like putting the fire out.

**The Astaxanthin effect is greater in the body.** While it's important to measure an antioxidant's strength in a laboratory, what's even more important is how it behaves in the body. Four important points to consider are whether the antioxidant can:

- Cross the blood-retinal barrier to bring antioxidant protection to the eyes. Astaxanthin – yes.
- Protect both fat-soluble and water-soluble parts of the cell, meaning does it get into the parts of the cell that are most vulnerable to damage? Astaxanthin – yes.
- Get into and protect muscle tissue. Astaxanthin – yes.
- Under certain conditions, become a pro-oxidant instead of an antioxidant and actually cause oxidation and subsequent cell damage. Astaxanthin – no!

An example of an antioxidant becoming a pro-oxidant is ingestion of excess iron.

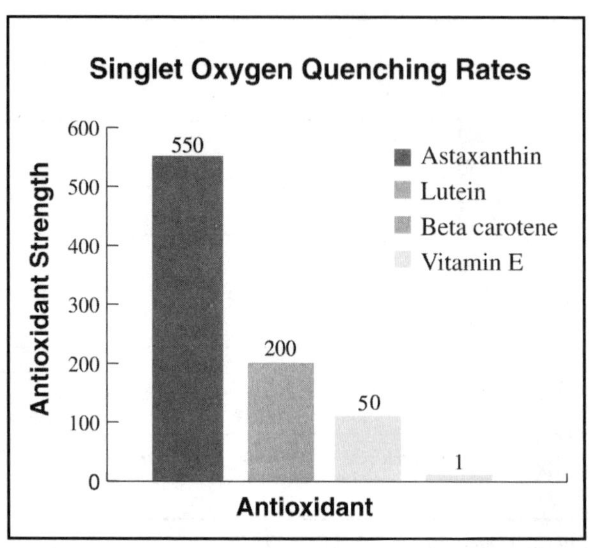


*N. Shimidzu, M. Goto and W. Miki 1996*

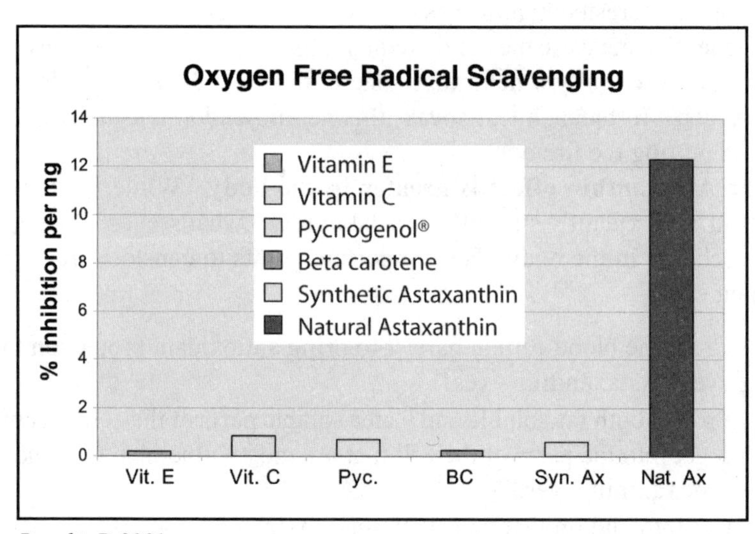


*Bagchi, D 2001*

You'll notice that the above graphs show different values according to different tests that are used. The three most popular tests are the singlet oxygen quenching test, the oxygen free radical scavenging test, and the ORAC test (oxygen radical absorbing capacity). For example, in the test measuring singlet oxygen quenching, Astaxanthin proved to be 550 times stronger than vitamin E. In the test measuring free radical scavenging, Astaxanthin was twelve times stronger than vitamin E. This is why it can be misleading to rely on a single test to measure an antioxidant's strength and why the popular ORAC test (not shown) is *not* the only one to consider. It's interesting that in all three antioxidant tests, Astaxanthin scored highly above all the rest. In the ORAC test, Astaxanthin proved to be much more powerful than other common antioxidants.

## LEARN THE CELLULAR SECRET TO HEALTH

A medical truism is the body is only as healthy as each cell in it. The healthier the cell, the healthier the whole body – that is, the whole is only as healthy as the sum of its parts. Since there are around thirteen trillion cells in the body, that's a lot of parts to keep healthy. Here's how Astaxanthin keeps cells healthy.

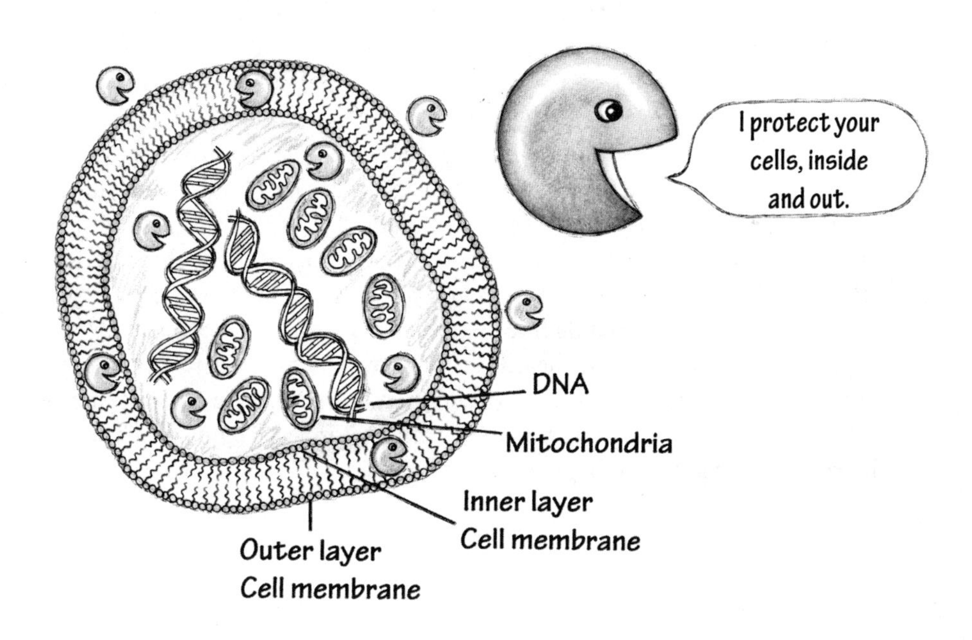

Three areas of the cell are most vulnerable to oxidative damage:

- The cell membrane, which is the outer lipid or fat layer of the cell
- The DNA, or genetic code within the cell, that dictates how correctly the cell replicates itself when it needs to multiply, grow, or replace itself when it wears out
- The *mitochondria*, which are like tiny batteries within the cell producing the energy the cell needs to do its work

Remember the principle of tissue wear and tear: The hardest-working tissues produce the most oxidants and need the most antioxidant protection. All three of these areas of the cell are hardworking and need the most antioxidants.

The cell membrane is one of the most marvelous structures in the entire body. Picture it like a flexible covering, or skin, that encompasses all the cell contents, sort of like the covering of a ball.

## PROTECT YOUR CELL MEMBRANES

One of the microscopic secrets of wellness that few folks appreciate is how important a healthy cell membrane is to overall health. Let's take a trip across this marvelous membrane to understand how it works and how we can keep it healthy.

**Leaky cell membranes.** Each cell in your body is like your body's energy and nutrient bank. There is a lot going on inside there that needs protection by the membrane. The cell membrane has two major functions: (1) It transports nutrients from the outside to the inside of the cell, that is, from the bloodstream into the cell, and then excretes energy waste products back out. (2) It protects the vital cellular contents from leaking out. In fact, a new term used by doctors studying the cellular basis of disease is leaky cell membrane disease. You probably have no idea what that means, but it sounds like something you don't want to have.

So a microscopic secret to health is to have healthy cell membranes that effectively transport and protect. Healthy cell membranes are the basis of internal health, like when you go to an internal medicine doctor. Taking care of your cell health is where internal health begins.

**The sick cell syndrome.** The cell membrane is a marvelous structure. It's composed of roughly half protein and half fat. One of the most metabolically magnificent construction projects within the body is how the fat molecules within the cell membrane line up to both protect and transport. Remember, oil and water don't mix, which is why the membrane contains a lot of fat, allowing the cell to sit next to a stream of liquid (blood vessels and tissue fluids) without dissolving. Drop an all-sugar piece of candy into a glass of water and see how quickly it dissolves. Now take a chocolate chip and see how it dissolves more slowly (it contains some fat). Now if you would take that chocolate chip and remove all the sugar, add protein and certain fats, it wouldn't dissolve at all. It's a good thing our cell membranes are designed as they are, otherwise our bodies would quickly "dissolve." Many illnesses like chronic fatigue syndrome and fibromyalgia are thought to biochemically be a reflection of the "sick cell syndrome."

## ASTAXANTHIN PROTECTS THE CELL MEMBRANE

Fat is the tissue most vulnerable to oxidation, that disease-producing "shun" word. Here's how to protect it. I remember that smelly filet of fatty fish that turned rancid (oxidized) after I mistakenly left it on top of our freezer overnight. To protect this precious membrane fat, you need the precious pink nutrient, Astaxanthin. If Astaxanthin could talk, she would say: "I protect your fat." Here's why membrane fat and Astaxanthin are best buddies.

Astaxanthin protects cell membranes better than any other antioxidant, including vitamin C and beta carotene. Fat-molecule membranes, called cellular membrane lipids, line up vertically like a picket fence, enclosing the precious contents of the hardworking, growing, and reproducing animals inside, the cell contents. The membrane, or fence, lets in what the animals need, keeps out what they don't need, and transports back out, shall we say, the "stuff" that builds up within the cellular pasture.

What makes Astaxanthin such a special cell-membrane protector? It enjoys a double biochemical property: *lipophilic*, meaning it loves fatty tissue, and *hydrophilic*, meaning it also loves water. This enables it to work in tissue that contains both fat and water – the structure of the cell. *Both* surfaces of a cell's fatty membrane need protection: the surface that is in contact with the fluids outside the cell, which keeps it from dissolving into the bloodstream, and the surface that's in contact with the fluid inside the cell. The antioxidant vitamin C helps protect the outside fat molecules because the outside surface is water-soluble, but it can't protect the inside, or fat-soluble layer of the cell. So, it only does half the job. Beta carotene does just the reverse. Because it can penetrate fat, beta carotene protects the fat inside the cell membrane from oxidation, but it doesn't protect the water soluble surface. Astaxanthin does both. Because of its unique biochemical structure, it spans the entire membrane to protect both layers – the lipophilic and the hydrophilic. Most antioxidants are soluble in either fat or water, but not both. Astaxanthin is a unique antioxidant in that it is both – fat and water soluble – being able to protect both the outside and inside of the cell. As you will learn throughout this book, some of the most vital tissues in your body are fat: the fatty layer in cell membranes, the fatty layer that wraps around nerve cells, the retinal tissue of the eye, and, of course, the skin you see and feel. Like a security guard, Astaxanthin easily penetrates fatty tissue and helps protect against foreign invaders like germs and oxidants that could get into the cell membrane and damage it.

If the cell membrane could talk it would say to Astaxanthin "We're made for each other!"

## ASTAXANTHIN PROTECTS THE INSIDE OF THE CELL

Astaxanthin is aptly called the outside/inside cell health protector. The micro factory of a cell works 24/7, producing oxidants (waste) that are neutralized by the cell's own antioxidants. One of the most well-known oxidants the cell produces is *superoxide*, which is neutralized by a natural cellular enzyme called *superoxide dismutase*, better known as SOD. But, as cell biologists say, "accidents happen" within the hardworking cellular factory. Some of these oxidants escape into the cellular machinery before the SOD police can grab them, bind them, and prevent them from doing harm, such as oxidizing or rusting the cells.

Another protective property of Astaxanthin is its *double bonds*, the biochemical structure that makes a molecule a powerful antioxidant. Think 007, Bond – double bonds. Double bonds mean they have an extra parking space for any free wayward electrons (oxidants) to park. They act like magnets to grab onto the oxidants and get them out of circulation.

## ASTAXANTHIN PROTECTS THE MIGHTY MITOCHONDRIA

The new science on aging tells us that the micro factories within each cell – called mitochondria – rust and wear out too fast. Mitochondria are some of the hardest-working molecules in the body. Remember, the hardest-working molecules generate the most oxidants and therefore need the most antioxidant protection. Preserving the life of the cellular machinery is the main key to longevity. Healthy aging, therefore, is keeping the cellular machinery protected from rust. This is what powerful antioxidants, like Astaxanthin, do to slow down cellular aging. Take home message: The younger each cell, the younger your whole body.

# CHAPTER 3

# THE HEAD-TO-TOE HEALTH EFFECTS OF ASTAXANTHIN

Let's take a trip through your body to see the Astaxanthin effects on your most vital organs. Since cardiovascular disease is the number one cause of death throughout the world, let's start with the heart.

## BE HEART SMART

Deep within your cardiovascular system lies one of the top secrets to health. Its discovery won the Nobel Prize.

**A night to remember.** As I was searching for the secrets to health, I was fortunate to learn from wise scientist friends and the health discoveries they made. As a physician, I believe that our bodies are designed to make most of the internal medicines we need. But where in the body is this internal pharmacy and what medicines does it make? The answer to this secret to health came one night at our home when Nobel laureate, Dr. Lou Ignarro, was our guest for dinner. Dr. Lou taught me where this pharmacy was and what it did (Ignarro, 2005; Sears, 2010).

**Save your silver lining!** You're about to learn what "silver lining" means to your health, where it is in your body, and how you can keep it healthy. A medical truism is: "You're only as healthy as your blood vessels," because the health of every organ depends on the nourishment it gets through its blood vessels. The lining of your blood vessels, known medically as the endothelium, is what I call the "silver lining" because it is the key to health and longevity. Your endothelium is a one-cell-layer tissue that lines the inside of your blood vessels. Think of it like wallpaper on the walls of your vessels. Yet, unlike wallpaper, it doesn't just sit there and look pretty. It does something. In fact, your endothelium is the largest hormone-producing tissue in your body. If you were to open up all your blood vessels and lay them out flat, your endothelium would occupy a surface area larger than several tennis courts. The endothelium is one of your most important systems, yet many people don't fully appreciate what it is and what it does.

Picture your endothelial lining containing millions of microscopic "medicine bottles" (biochemically called secretory cells). A healthy person who eats lots of antioxidants like fruits, vegetables, seafood, and Astaxanthin enjoys less build-up of sticky stuff (oxidants) on the endothelium. So the endothelial "pharmacy" stays open: *Wellness*.

Endothelial dysfunction occurs when an antioxidant deficiency allows sticky stuff to build up, preventing the medicine bottles from opening. So the endothelial "pharmacy" is closed: *Illness*.

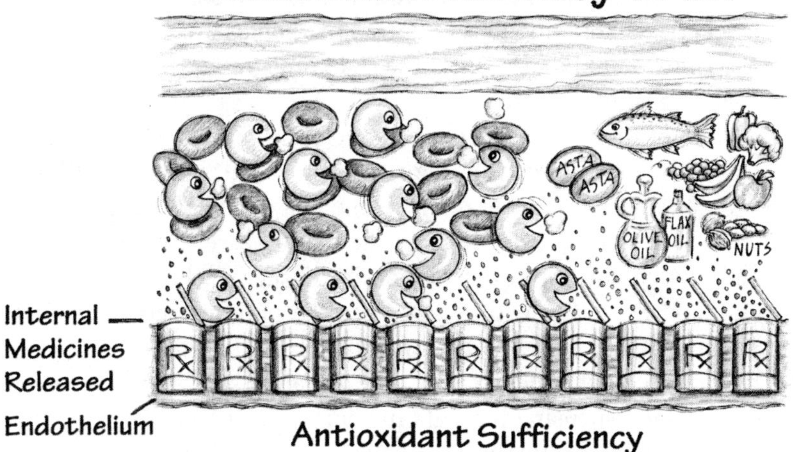


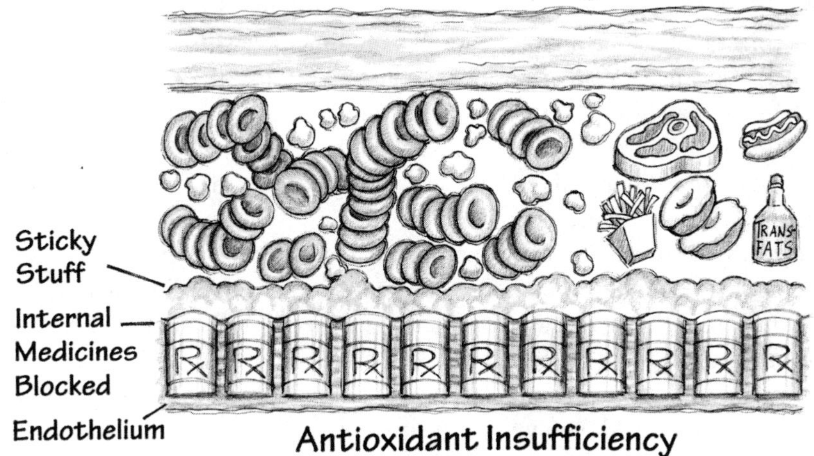

During our "scientist and simpleton" dialogue, as Lou was explaining his research in heavy biochemical terms, I was drawing my medicine bottle explanation. That's how I came up with the teaching tool of the endothelium being your personal pharmacy within and how the build-up of sticky stuff keeps the pharmacy closed.

**Care for your personal pharmacy.** Consider your endothelium as your own internal drugstore churning out natural internal medicines 24/7. Yet, unlike the pharmaceuticals you buy, these internal medicines are custom-made just for you, released into your bloodstream at the right time, in the right dose, and without any side effects, and they're free.

The cells of the endothelium are like closely placed tiles on a countertop. But again, these tiles don't just sit there, they do something. Each tile is its own endocrine organ, like millions of microscopic medicine bottles. The endothelium is like the chemical command center of the blood vessels, telling them when to open up and deliver more blood to hardworking tissues, such as when the intestines need more blood flow after eating and when the muscles need more blood flow during exercise. In some ways, the endothelium is like having your very own internal medicine doctor inside your body, constantly sensing your medical needs and responding by dispensing the right medicine. In chemical speak, these small molecules and proteins within the endothelial cells ("medicine bottles") are known as biologically active substances.

**What the "medicines" in the silver lining are.** There are around 23 known chemicals, or "medicines," within those tiny medicine bottles in your endothelium. There are medicines to lower the "highs" from high blood pressure and high blood clotting (like vasodilators and blood thinners); medicines to elevate the "lows" and mellow your moods (like antidepressants); and medicines to heal your hurts (like anti-inflammatories). You will notice we just mentioned the medicines that many people already have stocked in their medicine cabinet.

Think of these medicine bottles inside the endothelium as millions of tiny pharmacists. The better you feed and care for these pharmacists within, the better they dispense the medicines you need.

**Healthy endothelial function is the key to good health.** "Endothelial dysfunction," a relative newcomer to the growing list of cardiovascular diseases, is really how most heart disease begins. "Endothelial dysfunction" is the medical term for a lot of sticky stuff collecting on the silver lining that, in effect, prevents the natural internal medicine bottles from opening. Here's the simplest explanation for cardiovascular health you've ever heard:

Keep sticky fats, sticky inflammation products, and other sticky stuff off your vascular highways. This graphic explanation of cardiovascular disease is illustrated on Pages 28-29.

If I were asked to describe the one goal or change a person could make to enjoy healthy organs and longevity it would be: Keep your endothelium healthy. Or even more basic: Keep the sticky stuff off your endothelium. Astaxanthin helps keep the sticky stuff off your endothelium. As I was putting together my own health plan, I realized that one of my top goals for health and longevity was to keep my endothelial pharmacy open and releasing medicines as long as I can. Since every organ in the body is only as healthy as the blood vessels nourishing it, total body health begins with a healthy endothelium.

## FIVE WAYS ASTAXANTHIN CAN HELP YOUR HEART

Most heart disease is caused by the build-up of sticky stuff (oxidation and inflammation) in the blood vessels. Astaxanthin is a powerful antioxidant and anti-inflammatory. A healthful fit!

1. **Astaxanthin reduces the deposit of sticky stuff on the endothelium.** Astaxanthin, as an antioxidant, binds the oxidants (sticky stuff) before they have a chance to stick to the endothelium. Having a healthy silver lining is like painting a highway with a non-stick surface so that the cars (blood cells) can keep moving quickly and not cause a traffic jam (like cells sticking to the sides of an artery). Think clots, stroke, coronary thrombosis.

An interesting study showed that Astaxanthin decreases the ability of inflammatory cells (sticky stuff) to infiltrate the plaques of arteries in experimental animals, a double process leading to more build-up of plaque that eventually can clog the artery. "Stiff" and "sticky" are the two bad words for cardiovascular health. As sticky stuff gradually builds up in the walls of the arteries, the arteries get stiff, causing hardening of the arteries, the most common cause of high blood pressure. With less stiff and sticky stuff in their way, the arterial walls stay appropriately softer, helping keep blood pressure normal. (Hussein, 2005; Monroy-Ruiz, 2011)

While the blood pressure-regulating effect of Astaxanthin is probably due to its antioxidant/anti-inflammatory effects, when researchers examined the blood vessels of Astaxanthin-supplemented hypertensive rats, they discovered that the aorta, the major blood vessel coming from the heart, was less thick (i.e., less hardening of the arteries) than the aorta of the rats that didn't get the supplement. These researchers concluded that since high blood pressure was due to endothelial dysfunction, or secretion of vasoconstrictors (biochemicals that narrow the vessels and make them stiffer), Astaxanthin helped regulate these substances. (Hussein, 2006)

2. **Astaxanthin can help prevent heart failure from high blood pressure.** The most common cause of heart failure is endothelial dysfunction. The blood vessels, because they are so stiff with sticky stuff accumulation on their walls, can't relax. As a result, the pump at the other end, the heart muscle, has to work harder to pump blood through them. If they were wider, softer, and more relaxed, the heart wouldn't have to work so hard. Try blowing up a balloon. Initially, when the balloon is stiff and partially collapsed, it takes a lot of blow-pressure to open it up. Once the balloon is less stiff and more relaxed, it takes much less pressure to complete the inflation. That's what your heart muscle faces with every beat. If there is high pressure at one end,

the heart muscle at the other end is going to eventually fail before its time.

Here's what science says about the Astaxanthin effect on high blood pressure:

> Research in experimental animals shows that Astaxanthin can help alleviate high blood pressure and increase the strength of heart-muscle contraction to help the heart maintain normal blood flow, especially after injury, such as after a heart attack. (Gross, 2005; Nakao, 2010)

A study on adults with metabolic syndrome, a precursor to heart attack, stroke, and diabetes, showed that Astaxanthin helped reduce high blood pressure by reducing arterial stiffness. (Satoh, 2009; Fassett, 2008)

Japanese researchers discovered that supplementation with Astaxanthin for fourteen days resulted in a significant decrease in blood pressure in hypertensive rats, but showed no decrease in rats with normal blood pressure. They also showed that stroke-prone rats that were fed Astaxanthin for five weeks had a delayed incidence of strokes, in addition to a decrease in their blood pressure. This study found that rats fed Astaxanthin had measurable differences in their cardiovascular "plumbing." The two major findings were that Astaxanthin decreased elastic bands in the aorta (sticky tissue that causes arterial stiffness or hardening of the arteries) and decreased the thickness of the arterial walls. The general conclusion was that Astaxanthin helped relax and keep the blood vessels from "hardening." (Hussein, 2005, 2006)

3. **Astaxanthin helps reduce heart attack effects.** A fascinating study in experimental animals showed that those animals who had a heart attack, or damage to the heart muscle, and were given Astaxanthin, showed less damage and dysfunction of the heart muscle. Researchers concluded that Astaxanthin helps neutralize the damaging effects of the oxidants produced in the oxygen-deprived tissue during a heart attack. (Lauver, 2005; Curec, 2010)

Another study found that Astaxanthin-fed mice that ran on a treadmill until exhaustion suffered less heart damage than mice that were similarly exercised without Astaxanthin supplementation. On autopsy examination, researchers found Astaxanthin concentrated in the mice's hearts, leading them to conclude that Astaxanthin can decrease exercise-induced damage to the heart and possibly other muscles. (Aoi, 2003) In a study of human volunteers, participants supplemented with 6 milligrams of Astaxanthin per day for only ten days showed a significant improvement in blood flow. (Miyawaki, 2008)

4. **Astaxanthin decreases inflammatory markers for cardiovascular disease.** Human studies showed that natural Hawaiian Astaxanthin decreased the blood level of CRP (C-reactive protein), a biochemical that goes up in the bloodstream when there is excessive inflammation in the body. Elevated CRP is one of the markers that doctors measure in assessing a person's risk for cardiovascular disease. (Spiller, 2006)

5. **Astaxanthin helps improve the blood lipid profile.** Studies show that Natural Astaxanthin can decrease possible artery-clogging fats, such as oxidized LDL and triglycerides, and increase artery-cleansing fats, such as HDL.

Here's where science validates common sense. When healthy blood fats, such as cholesterol, "oxidize" they become bad cholesterol because they stick more to the tissues, such as the lining of the blood vessels. Tests show that Astaxanthin decreases the ability of LDL cholesterol to become oxidized. (Awamoto, 2000)

A 2010 study from Japan showed that humans who took Astaxanthin supplements showed *an improved lipid profile*, namely decreased triglycerides (sticky fats) and increased HDL-cholesterol (non-sticky fats). Blood levels of the hormone adiponectin were also higher in the people who took Astaxanthin. Adiponectin is a newly discovered natural hormone that helps regulate blood sugar and blood fat levels. A deficiency of this hormone is being implicated as contributing to type II diabetes and metabolic syndrome. Optimal results were found at 12 milligrams of Astaxanthin per day. (Yoshida, 2010)

Not only can Astaxanthin protect the good fats from becoming bad fats, which would cause them to oxidize and stick to the sides of arteries, but a further study of humans taking Astaxanthin supplements showed that Astaxanthin also decreased total cholesterol and LDL cholesterol by 17 percent and decreased triglycerides (other sticky fats) by an average of 24 percent. (Trimeks, 2003)

A study by Harvard researchers revealed another clue to how Astaxanthin helps hearts. Remember the anti-inflammatory drug Vioxx® that was pulled because it increased the risk of heart disease? The researchers discovered that Vioxx harmed hearts by increasing the susceptibility of heart cell lipid membranes and LDL cholesterol to oxidation. In other words, it behaved like a *pro-oxidant.* But, the researchers went on to find that Astaxanthin, a powerful antioxidant, blocked the oxidation effect of Vioxx during experiments. Simply put, Vioxx attacked tissue fats; Astaxanthin protected tissue fats. (Mason, 2006)

In fact, cardiologists now conclude that most of the cardiovascular benefits of Astaxanthin are due to its antioxidant and anti-inflammatory effects. In a nutshell, it decreases oxidative stress on the heart and blood vessels. (Pashkow, 2008)

Good science and good sense usually go together. Most cardiovascular disease is caused by inflammation and oxidation. Astaxanthin is both a powerful anti-inflammatory and antioxidant. Go red!

## HOW ASTAXANTHIN PROTECTS BRAIN HEALTH

Remember one of the top health tips: Because the hardest-working tissues produce the most oxidants, they need the most antioxidants. Three unique features of the brain make it a high-need tissue for antioxidants like Astaxanthin.

1. The brain is the hardest-working tissue of the body. It uses 25 percent of the carbohydrate energy you eat and burns 20 percent of the oxygen you breathe. Yet, your brain makes up only about 2 percent of your body weight.

2. The brain is 60 percent fat, and fatty tissue is the most vulnerable to oxidation or the excessive wear and tear known as "oxidative stress," even more so than high-protein tissue like muscle. You might say that the brain, especially as we age, gets stressed out.

3. The blood-brain barrier (BBB) screens out chemicals, called neurotoxins, that may harm brain tissue. However, the BBB often weakens with age. The aging brain is also less able to mute the effects of high-stress hormones, a condition called *gluconeurotoxicity.* And, the aging brain is less able to repair itself and grow new brain tissue.

Because your brain is such an important – and vulnerable – organ, a new term has been coined for nutrients that protect nerve tissue, *neuroprotectants.*

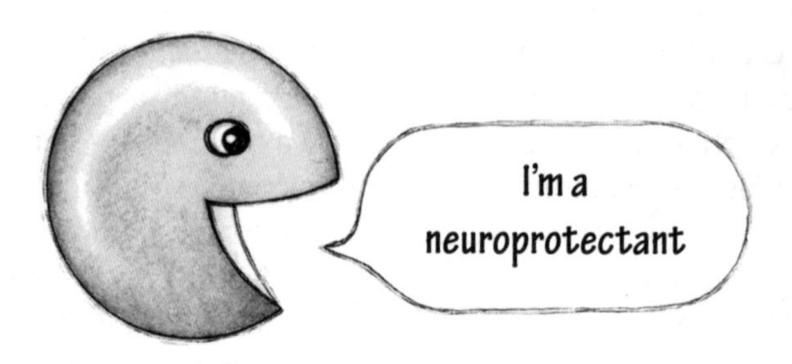


For obvious reasons, before supplements or drugs are used in humans, they are tested on experimental animals to see if they are safe and effective. The Astaxanthin effect on nerve tissue protection and repair has passed the animal testing phase, and we are eagerly awaiting the results of more human testing. As much as possible, researchers try to recreate in animals the tissue damage or disease found in humans, such as insufficient blood supply from a stroke or oxidative stress from inflammation, the two most common causes of nerve tissue diseases in humans. Here's a summary of what science says about Astaxanthin as a neuroprotectant:

**Slows age-related cognitive decline.** Ten seniors with age-related forgetfulness received 12 milligrams (mg) of Natural Astaxanthin daily for twelve weeks. Astaxanthin was shown to be effective for slowing their age-related cognitive and psycho-motor-function decline (Satoh, 2009).

**Lessens oxidative stress to the brain.** After twelve weeks of taking 6 or 12 mg daily of Natural Astaxanthin, human volunteers were found to have a decreased blood level of oxidants called phospholipid hydroperoxides (sticky toxic stuff that damages brain fatty tissue), which accumulate in the blood and nerve tissue of people suffering from dementia (Nakagawa, 2011). A study from the Departments of Surgery and Pharmacology at the University of Pittsburgh School of Medicine set up a mouse model for Parkinson's disease and found that Astaxanthin offers neuroprotection against oxidative stress caused by cerebral ischemia, or insufficient blood flow to nerve tissue (Lee, 2011). Researchers at the Nagoya University in Japan showed that human brain cells (in a laboratory dish) damaged by oxidation showed less damage when they had been pretreated with Astaxanthin, leading the researchers to conclude that Astaxanthin is a "natural brain food" for oxidative stress-induced damage to brain cells (Liu, 2009).

**Protects against brain damage and accelerates nerve tissue repair**. Brain tissue that was experimentally deprived of blood flow to mimic a stroke showed less damage when the tissue was pretreated with Astaxanthin five hours prior, and again one hour before, the blood supply to the tissues was reduced (Lu, 2010).

Using high doses of antioxidants in the treatment of traumatic brain injuries (TBI) is currently being researched and used by the military for combat injuries (Nutrition for WarFighters Conference, 2007). Again, science and common sense go together. If injured brain tissue builds up oxidants and inflammation, then treatment with a powerful antioxidant/anti-inflammatory should help.

Another laboratory study showed that damaged nerve tissue pretreated with Astaxanthin prolonged the survival of the damaged nerve cells. In laboratory speak, it protected them from "induced oxidative cell death" (Wang, 2010; Chan, 2009). A further study demonstrated the neuroprotective effects of Astaxanthin pretreatment of mice brain cells that were damaged from high doses of an oxidant (Kim, 2009).

Damaged nerve tissue, such as following a traumatic brain injury or a stroke, can repair itself by producing a type of neurological stem cell called a *neuroprogenitor cell* (NPC). A fascinating university study from Korea showed that pretreatment with Astaxanthin increased the proliferation of NPCs in experimentally damaged nerve tissue in mice. Researchers think that when a cell is damaged it sends out signals to other cells that it's in need of repair. The NPCs are like front-line medics summoned to the battle scene to treat traumatic brain injuries before they become fatal (Kim, 2010).

Researchers at the National Institutes of Health showed that rats that suffered an induced stroke from occlusion of a major brain artery, which led to blood-deprived brain tissue, had less damage to their brain tissue when the rats were pretreated with Astaxanthin. The researchers concluded that tissue in the *brain* damaged by ischemic injuries (strokes) may respond to treatment the same way that tissue in the *heart* responds after it has been damaged during a heart attack, as mentioned on Page 33 (Shen, 2009).

Since high blood pressure is the cause of strokes and brain damage, scientists at the International Research Center for Traditional Medicine in Japan experimented on rats with high blood pressure by giving them Astaxanthin. They found that experimental animals that had been pretreated with Astaxanthin had a lower incidence of stroke. They also were able to induce a stroke in mice by blocking the carotid artery. This recreated the human condition of when the carotid artery gets blocked from the build-up of plaque, depriving the brain of blood flow. The mice fed Astaxanthin an hour before the induced stroke showed less brain tissue injury than the mice not given Astaxanthin, further validating Astaxanthin's neuroprotective effects (Hussein, 2005; Kudo, 2002).

### ASTAXANTHIN CROSSES THE BLOOD-RETINAL BARRIER AND ACTS AS A STRONG RETINAL PROTECANT

A fascinating tribute to the architecture of the human body can be seen in the eyes which can be drastically affected by potential toxins in the bloodstream. The eyes are protected by a one-cell thick wrapping called the blood-retinal barrier (BRB). In studying Astaxanthin's effect on the eyes, researchers wondered if Astaxanthin had the ability to cross the BRB since many antioxidants and anti-inflammatories cannot. Way back in the 1940's, Drs. Rene Grangaud and Renee Massonet in their doctoral research at the University of Lyon in France were the first to discover that Astaxanthin does indeed cross the BRB and into eye tissue. By feeding laboratory rats Astaxanthin, they went on to prove that once Astaxanthin reaches the eyes, it acts as a strong retinal protectant (Grangaud, 1951 and Massonet, 1958). While scientists have not conclusively proven that Astaxanthin crosses the blood-brain barrier (BBB), it probably does as the previous studies on brain health effects seem to suggest.

## ASTAXANTHIN IS "SEE" FOOD

The retina of your eyes, the back part of your eyes that functions as a projection screen, is an extension of brain tissue. Therefore, what's good for the brain should be good for the eyes. Science agrees. And, like brain tissue, retinal tissue is one of the most fatty and hardest-working tissues in the body. Both brain cells and eye cells are what I call *high-need cells*, those hard-working (high-metabolic) cells that need more antioxidant protection. The retina and especially the sensitive macula on the retina function like a camera that captures the photo of, say, your beautiful baby or grandbaby. Their tissue is composed mainly of fat and a network of blood vessels which are highly affected by oxidation. In fact, retinal tissues have some of the highest metabolic rates of any tissue in the body. Because of its high rate of energy use, it generates lots of waste products, or oxidants, so it needs a powerful antioxidant. Let's focus on how Astaxanthin can help protect your vision.

Imagine your eye retina taking millions of snapshots in a millisecond, processing these into a composite photo of your smiling baby. Fortunately, the eyeball is one of the most examinable tissues to study. Using an ophthalmoscope to look through that little black opening in your iris, your doctor can examine the back of your eyeball – the retina – and observe how over time it responds to your healthy dietary changes, like putting more antioxidants into your diet. This is why many people, like myself, notice clearer vision after adding antioxidants to their diet.

A well-known xanthin – zeaxanthin – has long been recognized as an important nutrient for eye health. Studies have shown that zeaxanthin, found mainly in dark-green leafy vegetables, corn, peppers, and egg yolk is an essential nutrient that gets into the eye tissues, especially the lens and retina, to protect them against the aging effects of excessive ultraviolet light radiation. But there is another xanthin that gets into eye tissue – Astaxanthin.

Zeaxanthin and its buddy lutein, both carotenoids, protect macular pigments in the retina necessary for retinal health and clear vision. Zeaxanthin absorbs harmful blue light and reduces retinal oxidation and inflammation, which is why people who ingest more lutein and zeaxanthin have a lower incidence of age-related macular degeneration, the most common cause of vision loss. Until recently, lutein and zeaxanthin were thought to be the only carotenoids in our diet that could protect the lens and retina of the eye. Astaxanthin is the newcomer to the xanthin eye-protection family.

SCIENCE SAYS MORE ABOUT ASTAXANTHIN FOR EYE HEALTH

Scientists conclude that Astaxanthin acts as a "see food" by its powerful antioxidant and anti-inflammatory properties, working in tissues most vulnerable to these two effects. As an additional perk, Astaxanthin can increase blood flow to the retina, further protecting the eyes.

**Protects tired eyes.** Eye strain or eye fatigue is a malady of modern VDT (video display terminal) life. Just as other tissues of the body get overworked, the eyes get tired. This is why antioxidant eye protection is even more important in today's high-tech living. Researchers in Japan found that people who spend a lot of time at VDTs and took 5 mg of Astaxanthin per day for four weeks reported a 46 percent reduction in eye strain and a higher accommodation amplitude, that is, the ability of the lens to properly focus. A lens that is better able to focus is less likely to get fatigued. Other studies using doses from 4 to 12 mg a day found similar improvements to eye fatigue, as well as less eye soreness, less dryness, and less blurred vision (Nakamura, 2004; Nita, 2005; Shiratori, 2005; Nagaki, 2002 and 2006; Yasunori, 2006). A study showed that heavy VDT users who were pretreated with Natural Astaxanthin as a preventive medicine recovered from their eye fatigue more quickly than those not given Astaxanthin (Takahashi, 2005).

**Helps you see better.** Several studies by researchers in Japan have shown that people who took Astaxanthin in doses ranging from 4 to 12 mg a day showed improved visual acuity (the ability to see fine detail) and depth perception (Sawaki, 2002; Nakamura, 2004).

### The More Blood Flow to the Tissues, the Greater the Astaxanthin Effect

More blood supply is needed to protect, nourish, and heal hard-working tissues like the retina. In a fascinating study, Japanese researchers measured retinal capillary blood flow in 18 volunteers who took oral Astaxanthin, 6 mg a day, for four weeks, compared with a placebo group. After four weeks of Astaxanthin supplementation, retinal capillary blood flow was significantly higher in both eyes, but unchanged in the placebo group. (Yasunori, 2005). The increased capillary blood flow sheds further light on Astaxanthin's retinal health effect.

**Helps keep eyes young.** Hard-working tissues like the lens and retina of the eye eventually wear out, causing two of the most common eye problems of aging: cataracts and ARMD (age-related macular degeneration), medical jargon for "the retina wears out." Because the lens and retina of the eye are two vital tissues affected by the normal wear and tear of aging, let's look at what science says about how Astaxanthin protects them. Around 70 percent of people over age 65 eventually get cataracts. Excessive wear and tear and inflammation of the lens makes it stiff and sticky, so it is less able to change shapes and accommodate the changing light of various visual images. A study from the Massachusetts College of Pharmacy and Health Sciences in Boston showed that Astaxanthin helped protect the formation of cataracts on the lenses of rats' eyes exposed to oxidants (Liao, 2009; Nakagima, 2008; Suzuki, 2005; Ohgami, 2003; Cort, 2010). Another study from the School of Pharmacy of Taipei Medical University in Taiwan showed that Astaxanthin protected the lenses of pigs' eyes from oxidation (Wu, 2006). One of the ways that macular degeneration develops is by overgrowth of blood vessels in the retina, a process called neovascularization. Researchers discovered that in mice whose retinas were stimulated to increase vascularization, those pretreated with Astaxanthin showed less abnormal growth of blood vessels (Izumi-Nagai, 2008).

Eye tissue, like brain tissue, is not only hyper-vulnerable to oxidation and inflammation, it's particularly resistant to repairing its cells and regenerating new ones. One of the reasons we believe antioxidant protection is so important for the eye is that the most sensitive cells of the retina called the macula, once damaged, are very difficult to repair and regrow. A medical lesson as true for the brain as it is for the eyes is: Prevention works better than repair.

## ASTAXANTHIN PROTECTS AGAINST INFLAMMATION

Most of a doctor's day is spent treating inflammatory diseases, especially the ABCD's: arthritis, bronchitis, colitis and cognivitis (Alzheimer's), and dermatitis. We currently have an epidemic of these "-itis" illnesses and, even more alarming, they are occurring in younger ages. On Page 12 you learned what inflammation is and what it does to the body. Now let's learn how Astaxanthin, one of nature's most powerful antioxidants, is also a powerful anti-inflammatory.

Over my 40 years in medical practice, I've come to the conclusion: The best medicines are found in nature. Dr. Mother Nature has the longest track record of producing the medicines with the two most important qualities: safety and efficacy. Mother Nature's medicines rarely get pulled off the shelf for being unsafe or not working. And Mother Nature doesn't continually change her mind with fad supplements that enter the market with a flurry and leave with a fizzle.

One of those natural medicines is Astaxanthin. Mother Nature has always produced antioxidants to balance the normal oxidation that occurs when living organisms burn fuel for energy. What Dr. Mother Nature hadn't counted on was that the modern living of the 21st century would produce another "shun" – inflammation. Much of the world's population is literally "on fire," or inflamed. Back to my sticky stuff explanation of illness you learned about on Page 11. In addition to causing oxidation, the sticky stuff that collects in tissues and makes them stiff and wear out – think arthritis – also causes inflammation.

You may wonder, "Why not just pop a pill?" Not so fast. Remember our two qualities of a good medicine? Safe and effective. While prescription anti-inflammatories are somewhat effective, many are not safe.

**Inflammatory balance is a key to good health.** On Page 15 you learned how the body uses biochemicals within your immune system to protect against illnesses and heal and repair. Health means your body is in inflammatory balance: The right balance of anti-inflammatory biochemicals are healing and repairing. If the wear and tear overwhelms the body's ability to repair, your body is inflamed. Or if your immune system's anti-inflammatory fighters overattack or get confused and attack the body's own tissues, the body will be out of inflammatory balance and an "-itis" illness, called *autoimmune disease*, will result. Inflammatory balance, or health, is the right balance of pro-inflammatory biochemicals and anti-inflammatory biochemicals. That is, "wear and tear equals repair." One of the most well-known pro-inflammatory biochemicals are a group of enzymes called *cyclooxygenases*, or COX for short. Medications, such as aspirin, ibuprofen, and the prescription medicine Celebrex, are called NSAIDS, or non-steroidal anti-inflammatory drugs. They are also called "COX inhibitors" because they partially suppress the pro-inflammatory COX enzymes.

Unlike the anti-inflammatories your body makes naturally, which are continuously adjusting their dose according to what's going on in your body, pharmaceuticals such as the NSAIDs are taken without knowing exactly the right amount to do the job. While NSAIDs may work on arthritis and some cardiovascular problems to block swelling, pain, and blood clotting, they can also *overreact* in the body. Instead of reducing the incidence of heart attack and stroke, they can actually cause these illnesses by over-thinning the blood, resulting in gastrointestinal bleeding and even cause blood clotting in coronary arteries. This is the reason Vioxx was taken off the market in 2004. Modern doctors stress to their patients the importance of eating an anti-inflammatory diet and having an anti-inflammatory lifestyle. Instead of taking drugs to suppress excess COX chemicals in your body, teach your body to make less of the COX chemicals in the first place. That's preventive medicine in a nutshell.

Pharmaceutical anti-inflammatories can be both hard on the heart by increasing blood clotting and irritating to the gut as well. Not only do these drugs increase gastrointestinal bleeding, they increase gastric acid and reduce the intestinal mucus lining that protects against too much acid, leading to gastrointestinal upsets such as ulcers.

To heal the hurts, people continue to reach out for pharmaceuticals. The problem with many of these over-the-counter and prescription drugs is that they heal one hurt, but cause another. Because of this growing concern over the safety of anti-inflammatory drugs, many persons who want to take charge of their health are searching for safer alternatives such as nutraceuticals. Astaxanthin may be one.

## WHAT SCIENCE SAYS ABOUT THE ANTI-INFLAMMATORY EFFECTS OF ASTAXANTHIN

There are two groups of COX pro-inflammatory enzymes, COX-1 and COX-2. Most of the harmful effects of pharmaceutical anti-inflammatories are caused by blocking too much COX-2, which may be one of the reasons they work faster than natural anti-inflammatories. On the contrary, natural anti-inflammatories, like Astaxanthin, block fewer COX-2 enzymes. They also suppress the overproduction of a broader range of pro-inflammatory biochemicals, namely prostaglandins, interleukins, and tumor necrosis factor alpha (TNF-a).

Because Astaxanthin works in a gentler, less concentrated way on a broader range of the body's inflammatory chemicals, it may be a slower, but safer, way to help heal your hurts. Biochemists who have studied Astaxanthin generally believe that Astaxanthin's anti-inflammatory properties are closely related to its powerful antioxidant activities. There seems to be a natural crossover between the antioxidant effect and the anti-inflammatory effect of Astaxanthin. Even though the following studies were conducted on small populations, the results are promising. Here's a list of what science says about the anti-inflammatory effects of Astaxanthin:

**Balances inflammatory chemicals**. Researchers studied rats and mice that had experimentally induced inflammation and then treated them with Astaxanthin. The blood levels of pro-inflammatory chemicals, such as prostaglandins, TNF, and interleukins, were then measured. The results showed that the Astaxanthin-protected animals had lower levels of inflammatory biochemicals than the animals not given Astaxanthin. Medically speaking, the animals showed a decrease in the blood level of "inflammatory markers." (Lee, 2003; Ohgami, 2003)

Even though Astaxanthin blocks the COX-2 pro-inflammatory enzymes, it does so in a much gentler way than pharmaceutical anti-inflammatories. For example, Celebrex was over 300 times stronger in COX-2 inhibition than Natural Astaxanthin. The ratio of COX-1 to COX-2 inhibition was 78:1 for Celebrex, whereas for Natural Astaxanthin it was only 1:1. Again, this gentler process may be one of the reasons Natural Astaxanthin is safer than many of the pharmaceuticals.

One of the most common markers indicating the presence of inflammation is called CRP, or C-reactive protein. CRP is produced in the liver and released into the bloodstream when the body is fighting inflammation. CRP does not in itself cause inflammation, but its presence alerts the doctor that something upsetting is going on in the body and needs to be diagnosed.

A 2006 human clinical study conducted by The Health Research and Studies Center in Los Altos, California, studied 25 people for eight weeks. Sixteen people were given natural Hawaiian Astaxanthin and 9 received a placebo. The group given Astaxanthin experienced a 20 percent reduction in CRP levels in just eight weeks, whereas the placebo group had an increase in CRP levels (Spiller, 2006).

**Helps heal the hurts of workplace injuries.** Many athletes suffer "-itis" illnesses and the pain of inflammation from overuse of their muscles and joints. In fact, joint pain from inflammation is one of the most common reasons people use anti-inflammatory drugs. Here are some studies showing how Astaxanthin helped "-itis" illnesses in the category of "overuse injuries."

**Carpal tunnel syndrome (repetitive stress injury, or RSI)**. With more and more people spending their days typing on computers, RSI is becoming one of the most debilitating workplace injuries. Because this condition doesn't always improve when people give their hands a rest, immobilize their injured wrists, or take pharmaceutical anti-inflammatories, people who suffer from RSI are turning to proven and safer alternative therapies. Researchers studied 20 people with RSI and divided them up into two groups: 13 people received 4 milligrams of Natural Astaxanthin three times a day, and 7 people received a placebo. Those given Natural Astaxanthin reported a 27 percent reduction in daytime pain after four weeks and a 41 percent reduction after eight weeks. The duration of their daytime pain decreased by 21 percent after four weeks and 36 percent after eight weeks (Nir and Spiller, 2002a).

**Helps heal the hurts of athletic injuries**. Tennis elbow, also known as lateral humeral epicondylitis – which sounds like something you don't want to get – is a common overuse injury among tennis players. In a 2006 study of 33 tennis players, 21 of whom received Natural Astaxanthin and 12 a placebo, researchers

found that after twelve weeks the Astaxanthin-treated group had a noticeable reduction in elbow pain and improved grip strength during their tennis playing (Spiller, 2006).

Astaxanthin was given to weight-lifting athletes to see if it could protect them from delayed onset muscular soreness (DOMS) in a study conducted at the Human Performance Laboratories of the University of Memphis. Nine participants were given a daily dose of 4 mg of Astaxanthin for three weeks prior to the weight-lifting session and during the twelve-day recovery phase. Five participants were given a placebo. Forty-eight hours after the weight-lifting session, the group taking Astaxanthin perceived less DOMS soreness than the group receiving the placebo (Fry, 2001).

**Helps heal rheumatoid arthritis**. Rheumatoid arthritis, a debilitating inflammatory disease, is becoming more and more common. It's in the category of an autoimmune disorder in which the sufferer's immune system attacks itself, as explained on Page 42. In an eight-week study, researchers studied 21 people with the disorder, 14 of whom received Natural Astaxanthin and 7 a placebo. Participants were evaluated on their degree of pain and their ability to perform daily activities. Pain scores for the group given Astaxanthin decreased by approximately 10 percent after four weeks and by more than 35 percent after eight weeks, whereas pain scores for the placebo group remained relatively constant. Those taking the Natural Astaxanthin rated their overall condition as having improved by approximately 15 percent after four weeks, and by over 40 percent after eight weeks. While this was a small study, it showed some promising effects of this powerful antioxidant on the disabilities accompanying rheumatoid arthritis. (Nir and Spiller, 2002b)

Inflammation is the painful effect of oxidation. Where you find one, you find the other. This is why a powerful antioxidant like Astaxanthin is also a helpful anti-inflammatory neutraceutical.

**A NUTRIENT FOR ALL AGES**

Every age suffers inflammation. Every age needs more inflammation protection. Young children have a young immune system which is why they get sick so often. Teens work their bodies hard during sports and suffer overuse injuries. Pregnant mothers really overwork their bodies. Young adults, a relative healthy period in life, should be in the "prevent mode" mindset, and seniors suffer ailments from tissues wearing out.

## ASTAXANTHIN HELPS BUILD A HEALTHY IMMUNE SYSTEM

In the previous section you learned how Astaxanthin helps balance inflammation. It also helps balance the body's immune system. We emphasize "balance" rather than "boosting" since, like inflammation, you want the body's immune system to react appropriately, not underreacting so that germs take over and you get sick, nor overreacting so that the immune system gets its signals mixed up and attacks its body's own tissue, hence the epidemic of auto-immune diseases. A weakened immune system is the underlying cause of many diseases and disabilities that daily keep kids out of school, adults out of work, and fill hospital beds.

Within the body's "armed forces," inflammation troops and immune system troops work together, sort of like the Army and the Marines. In fact, diseases of inflammatory imbalance are often triggered by immune system imbalances.

Circulating throughout your body 24/7 are microscopic soldiers that search out and destroy dangerous toxins, germs, cancer cells, and other things that if allowed to get a foothold within the body will make it sick. These immune system cells go by various names, but a few of the most popular are:

- *Macrophages* (literally "big eaters"). These are white blood cells that gobble up harmful foreign stuff before it can harm the body.
- *Natural killer cells* (NK cells). Also called B-cells and T-cells, these are some of the immune system's most diligent and heavy fighters that are called upon for big wars, such as killing cancer cells.

**Inside the laboratory of immune system researchers.** After immune system pioneers Drs. Chew and Park at Washington State University discovered that Astaxanthin boosts immunity in animals, they were able to show that it also increases the health of the immune system in humans. Here's a summary of what they discovered about how Astaxanthin helps keep the immune system healthy: To study the Astaxanthin effect on the immune system, they measured "markers of immune health," meaning biochemicals and germ-fighting cells that go up in a person's blood when the immune system is working at its best. Specifically, Drs. Park and Chew and their team of immune system researchers studied the Astaxanthin effect on healthy females of an average age of 21. The researchers gave them 2 to 8 mg of Astaxanthin a day for eight weeks in a double-blind, placebo-controlled study, the gold standard of research. After eight weeks of Astaxanthin supplementation, they found an overall improvement in the markers of immune system health: an increased level of NK cells and decreased damage to cellular DNA – an area of the cell most vulnerable to damage when the immune system is weakened. It's interesting that in this study a dose as low as 2 mg a day showed improvement in the immune markers. (Park, 2010)

## ASTAXANTHIN PROMOTES SKIN HEALTH

While Astaxanthin's effect on keeping skin healthier and looking younger has long been appreciated, a recent Hollywood buzz on Natural Astaxanthin has increased its popularity. More people are now learning about what dermatologists have long known: Natural Astaxanthin acts like an internal sunscreen. In 2011, England's second-largest newspaper, *The Daily Mail*, reported that Academy Award–winning actress Gwyneth Paltrow and supermodel Heidi Klum both were using Natural Hawaiian Astaxanthin to help beautify their skin. The article, titled "Extended Life Pill: Miracle Supplement Promises to Fight the Signs of Aging," lists these benefits for the skin:

- Lessens wrinkles and improves skin elasticity
- Reduces visible signs of ultraviolet damage
- Reduces risk of skin cancer

**Treat skin on the outside *and* inside.** In my medical practice, I teach it's not only what you put *onto* the skin to protect it, but also what you put *into* it, and that's where Astaxanthin shines. Skin ages, especially wrinkling and thinning, because of the collection of stiff and sticky stuff (oxidation) that infiltrate collagen, the protein and structural fibers of the skin. This weakens the skin's elasticity and causes it to sag. Since most skin damage is caused by oxidation it stands to reason that Natural Astaxanthin, a powerful antioxidant, would be good for skin health. Science agrees.

### SKIN DOCTORS PRAISE ASTAXANTHIN

Best-selling author and dermatologist Nicholas Perricone, M.D., in his book *The Perricone Promise: Look Younger, Live Longer in Three Easy Steps* praises the Astaxanthin effect on the skin. In his books and as a guest on "The Oprah Winfrey Show" he called Astaxanthin a wonderful anti-inflammatory and antioxidant that "gives you that beautiful, healthy glow." He also reported that it reduces age spots and wrinkles. Dr. Perricone attributes Astaxanthin's role as an internal beauty supplement to its antioxidant properties of primarily protecting the integrity of skin cell membranes exposed to oxidative-stress, mainly from sunlight and weather. (See more testimonials, Page 55).

**Astaxanthin acts like an internal sunscreen.** Sunburn is an inflammatory process. UV light inflames (or reddens) the skin and repeated overexposure to sunrays causes photo aging: wrinkled, blotchy, thinned, darkened skin. Since sunburn is inflammation and Astaxanthin is a powerful anti-inflammatory, it's a natural partner in skin health. To study the Astaxanthin effect on exposure to sun, a group of people were tested to see how much UV light was needed to redden their skin or cause a mild sunburn. They were then supplemented with 4 mg a day of Natural Hawaiian Astaxanthin for two weeks followed by a repeat of the skin-reddening test. A comparison between the pre- and post-skin-reddening scores showed that 4 mg of Astaxanthin per day increased the amount of time it took for UV radiation to redden the skin (Lorenz, 2002).

The skin is the body's largest organ, and yet in this study it took only two weeks for Natural Astaxanthin to work as an internal sunscreen. Because Astaxanthin has a cumulative effect in the body, building up in the organs over time, perhaps taking this powerful Astaxanthin for more than two weeks would have given the participants an even greater skin-protecting effect.

Since 1995 there have been many studies on animals showing how Astaxanthin prevents photo-aging from UV exposure. Yet only in the past decade has it been increasingly recognized as a skin-protectant in humans.

In 2006 a landmark human clinical study appeared in the journal *Carotenoid Research.* This placebo-controlled research used 49 healthy women with an average age of 47. The women were divided into two groups – one given placebos and the other supplemented with 4 mg a day of Natural Astaxanthin. At the end of six weeks, the women taking Astaxanthin reported so many improvements that Natural Astaxanthin was given the reputation of being an internal beauty pill. Here are some of the findings:

- In a self-assessment questionnaire, over 50 percent of those taking Natural Astaxanthin reported improvements in skin moisture content, roughness, fine lines and wrinkles, and elasticity.
- Dermatologists using instruments to measure skin health and beauty parameters found improvement in fine lines and wrinkles, elasticity and skin dryness, and moisture levels.
- Before and after photos showed visible improvements in fine lines, wrinkles, and elasticity (Yamashita, 2006).

This study was of interest since the conclusions were based not only on the observations of the women taking Astaxanthin but also on the dermatologists' examinations.

# CHAPTER 4

# QUESTIONS ABOUT ASTAXANTHIN YOU MAY HAVE

## BEST DOSE FOR YOU?

How much Astaxanthin should you take for the health effects you want? Surveying the scientific studies, it seems that most people would get the Astaxanthin effect they need by eating:

**4 – 12 mg of Natural Astaxanthin a day**

### WHERE TO FIND NATURAL ASTAXANTHIN

You can take it in supplements, eat it in food, or both.

**Natural Astaxanthin supplements.** The most scientifically researched Astaxanthin comes from Hawaii. I recommend Natural Hawaiian Astaxanthin because it's safety tested and comes from *Haematococcus* algae. Remember, fish eat algae to get this powerful pink antioxidant. Although human studies have demonstrated Natural Astaxanthin to be safe even in high doses, most studies have used 2 mg to 12 mg a day, the 12 mg reserved for persons suffering from severe inflammation.

**Why there are different dosages for different folks.** Bioavailability (how much your intestines absorb of what you eat) varies greatly among individuals. In consultation with your healthcare provider, you will need to experiment to find the right dose to produce the health effect you need. You may be a "high Astaxanthin absorber" and need a lower daily dose, or you may be a "low absorber" and need a higher dose. You may be very healthy and just want to keep your immune system in balance and enjoy a bit of preventive medicine. In this case, 2 to 4 mg a day may be enough. Or, you may be a metabolic mess and have a lot of inflammatory hurts and need 12 mg a day.

Take a tip from nature: Always take Astaxanthin supplements *with fats* at a snack or meal, or in a gel cap containing oil. Both nature and experiments have shown Astaxanthin is better absorbed when partnered with fat – think fatty sockeye salmon.

## SEAFOOD HIGHEST IN ASTAXANTHIN

**Best seafood sources are:**

Sockeye salmon: Averages 1 mg Natural Astaxanthin per ounce of sockeye. So to get 4 mg a day you would need to eat four ounces of sockeye salmon each day, or 1¾ pounds per week. I eat almost that much, but realistically few people will, so they will require a supplement. Other wild salmon species contain lesser amounts of Astaxanthin, around one-quarter of that found in sockeye. Usually the redder the fish, the more Astaxanthin it contains (Juruman, 1997). Other pink seafood such as lobster, crab, and shrimp have extremely small amounts of Astaxanthin.

Remember mother's wisdom: "You are what you eat!" Upgrade that to you are what the fish you eat eats.

**Fish are as "red" as they eat.** Because the content of Astaxanthin in seafood depends on how much that fish eats, there is a wide range of Astaxanthin concentration in seafood. For example, the average Astaxanthin concentration in wild Alaskan sockeye salmon is around *1 mg per ounce of fish*, around eight times the amount of Astaxanthin found in farmed Atlantic salmon.

| **Species** | **Astaxanthin Range** | **Astaxanthin Average** |
|---|---|---|
| Wild sockeye salmon | 30-58 mg/kg | 40.4 mg/kg |
| Wild Coho salmon | 9-28 mg/kg | 13.8 mg/kg |
| Wild pink salmon | 3-7 mg/kg | 5.4 mg/kg |
| Wild chum salmon | 1-8 mg/kg | 5.6 mg/kg |
| Wild Chinook king salmon | 1-22 mg/kg | 8.9 mg/kg |
| Atlantic salmon | 5-7 mg/kg | 5.3 mg/kg |
| **Average of all species** | | **13.2 mg/kg** |

## WHY NATURAL ASTAXANTHIN IS HEALTHIER THAN SYNTHETIC ASTAXANTHIN

For the best antioxidant effect, enjoy the synergistic effect. Antioxidants work best when eaten with an assortment of other antioxidants, capitalizing on a nutritional principle called *synergy*, meaning foods eaten together increase the health benefits of each of the separate foods, sort of like 1 + 1 = 3 or 4. This is why your mother told you to eat a *variety* of fruits and vegetables.

Not all Astaxanthins are created equal. Natural Astaxanthin is much more powerful than Synthetic Astaxanthin. Here's why. Natural Astaxanthin is grown in algae. Synthetic Astaxanthin is produced from petrochemicals. It seems unnatural to eat supplements made from the same oil you put into your car's engine. Also, Natural and Synthetic Astaxanthin molecules behave differently in the body because they are different molecules. While they may have the same chemical formula, there are three differences:

- The shape of the Synthetic Astaxanthin molecule is different from its natural counterpart.
- Natural Astaxanthin in nature is always paired with healthy fatty acids attached to either one or both ends of the Astaxanthin molecule.
- Natural Astaxanthin, unlike the synthetic stuff, contains a variety of other nutrients that work together like teammates. Dr. Mother Nature discovered this vital health principle of synergy millions of years ago, which is why antioxidants seldom exist by themselves in food.
- These biochemical perks make it behave much more healthfully in the body.

Remember, Dr. Mother Nature has had a much longer history of "manufacturing" the right nutrients in the right proportions than have chemical factories. For example, algae, namely *Haematococcus pluvialis*, the microalgae from which natural Hawaiian Astaxanthin is produced, accumulates other carotenoids as a survival mechanism. These additional carotenoids – all of which you frequently hear about in health circles: beta carotene, canthaxanthin, and lutein – work in synergy to make Natural Astaxanthin more powerful than the synthetic stuff.

Fish farmers have noticed that their fish fed Natural Astaxanthin are healthier than those fed Synthetic Astaxanthin, probably because it gets into tissues better than the cheaper synthetic stuff. Next time you're in a fish market, notice how much more pinkish-red the "natural" wild salmon are, compared to the pale-pink farmed ones.

Aquaculture (basically farming in water) experiments to make farmed fish healthier reveal that farmed Atlantic salmon grow better and survive longer when fed Natural Astaxanthin. As the level of Natural Astaxanthin in the tissue of baby salmon increased, their survival rates increased from 17 percent to a whopping 87 percent.

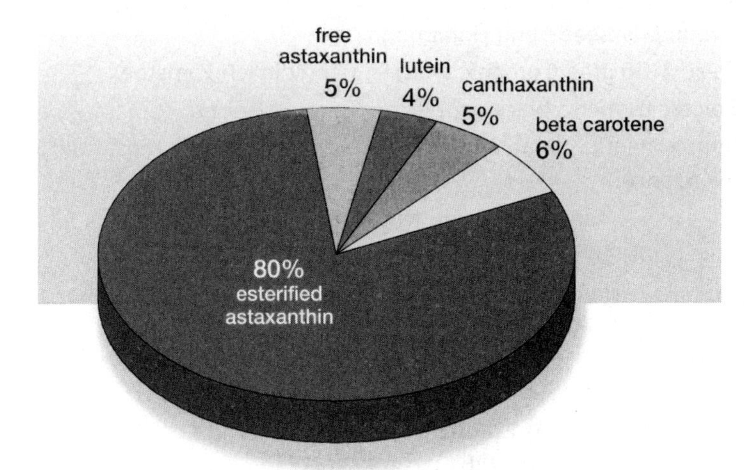


*Distribution of naturally-occurring carotenoids in Astaxanthin from Haematacoccus microalgae. Unlike Synthetic Astaxanthin, 80% of the Astaxanthin in the natural, algal-based form is "esterified" with a fatty acid molecule attached. Combined with naturally occurring carotenoids like lutein, beta carotene and canthaxanthin, this esterified Natural Astaxanthin is more potent and effective. It's just what Dr. Mother Nature ordered!*

## DR. BILL'S ANTIOXIDANT DIET

**Most of a doctor's day is spent treating antioxidant deficiency diseases. Here's the diet I prescribe, which I call the *5 S's*:**

- Seafood: primarily Alaskan salmon, tuna, sardines, and anchovies, all of which are high in anti-inflammatory nutrients.
- Smoothies: multiple dark-colored fruits, berries, organic yogurt, ground flaxseeds, and cinnamon.
- Salads: Go green; organic arugula, kale, spinach, tomatoes.
- Spices: turmeric, black pepper, garlic, rosemary, etc.
- Supplements – the three that I prescribe are the ones most supported by science:
  - — Natural Hawaiian Astaxanthin
  - — Omega-3 fish oils, primarily salmon oil from Alaska
  - — Juice Plus®, a whole-food-based concentration of many fruits and vegetables

These supplements fill in the gaps when you don't eat 12 ounces of seafood weekly and 9–12 servings of fruits and vegetables daily. And, let's add two more S's: homemade soups and stews, which, like smoothies and salads, enjoy the nutritional perk called synergy: Blend many nutrients together and each one becomes more healthful.

# CHAPTER 5

## ENJOY TRUSTED TESTIMONIALS FROM HAPPY ASTAXANTHIN TAKERS

Here are some real life stories from real Astaxanthin-supplementers – top scientists, top doctors, top athletes, and everyday folks who just want to feel better and perform better.

### PRAISE FROM A TOP PHARMACIST

• Pharmacist Suzy Cohen, a nationally syndicated columnist, TV personality, and author of the best-selling books *Drug Muggers, Diabetes Without Drugs*, and *The 24-Hour Pharmacist*, tells me why she takes Hawaiian Astaxanthin:

"Why do I take the antioxidant with a funny name? Because Hawaiian Astaxanthin is natural, and it has more cleaning power on the body than vitamin C, E, beta carotene, lutein, zeaxanthin, lycopene, pycnogenol, and alpha lipoic acid. For a good many years scientists have known that Natural Astaxanthin can prevent cataracts and macular degeneration, two common causes of blindness in this country. But in just the past few years Japanese researchers have found a fabulous new benefit for the powerful antioxidant. Just 6 milligrams of Astaxanthin a day can dramatically improve blood flow to the retina which nourishes the eyes and helps prevent eyestrain, which makes this dietary supplement a MUST for anyone with diabetes. I take Astaxanthin because it helps me with energy, and minor aches and pains. Plus, Astaxanthin makes your skin pretty – studies now prove it!"

### SELECTED BY A TOP SKIN DOCTOR

• Best-selling author and dermatologist Nicholas Perricone, M.D., in his book *The Perricone Promise: Look Younger, Live Longer in Three Easy Steps* praises the Astaxanthin effect on the skin. In his books and as a guest on "The Oprah Winfrey Show" he called Astaxanthin a wonderful anti-inflammatory and antioxidant that "gives you that beautiful, healthy glow." He also reported that it reduces age spots and wrinkles. Dr. Perricone attributes Astaxanthin's role as an internal beauty supplement to its antioxidant properties of primarily protecting the integrity of skin cell membranes exposed to oxidative-stress, mainly from sunlight and weather.

## NOTED BY A TOP NEUROSCIENTIST

- Paula C. Bickford, Ph.D., Senior Research Career Scientist, James A. Haley Veterans Hospital, Tampa, Florida; Professor, Department of Neurosurgery and Brain Repair, University of South Florida

"Astaxanthin has tremendous health benefits. Research on Astaxanthin has shown that it reduces markers of systemic inflammation and also reduces damage to DNA. We know from years of research that as we get older there is a background increase in the activity of cells that produce these markers of inflammation. The research links this background of inflammation to the fact that as we age there is a higher chance to develop neurodegenerative diseases and other immune related diseases. One way to help ourselves to feel younger and have more energy is to take Astaxanthin. I have added this to my regimen and feel a lot more energy. There are many hundreds of research studies supporting the benefits of Astaxanthin. Astaxanthin is the most potent carotenoid that has been identified and studied."

## TESTIMONIALS FROM TOP ATHLETES

### WHY ATHLETES LOVE ASTAXANTHIN

Athletes overuse their muscles, tendons, and joints. "Overuse injuries" account for much of the pain, soreness, and decreased performance that many athletes endure, especially following a marathon performance.

**THE TWO A's ARE A PERFECT FIT**

Simply put, heavily worked tissues produce excess oxidants and inflammation. Astaxanthin is a powerful antioxidant and anti-inflammatory. **Athletes and Astaxanthin are buddies.**

In my medical practice, I believe that good medicine makes good sense. An athlete's body makes excess oxidants, therefore an athlete needs to eat more antioxidants. You can do that!

Let's see what science says and read the personal testimonies athletes give. While many of these testimonies are from professional athletes, even amateurs who simply enjoy exercising their bodies may profit from Astaxanthin. In fact, most overuse injuries occur in amateurs whose muscles are not well trained for a sprint. It's common for these weekend athletes to wake up on Monday morning with sore muscles and joints and realize: "Oh, I overused my body."

- **Max Burdick** is an Ironman triathlete (swim for 2.4 miles, bike for 112 miles, and run for 26.2 miles). Max kept competing, but could never finish the triathlon because halfway through the bicycle stage his burning leg muscles would force him to stop. Then after supplementing with 8 mg of Hawaiian Astaxanthin daily he was able to finish the triathlon – at age 75!

- "My name is **Tim Marr** and I'm a professional triathlete from Honolulu. I first discovered Hawaiian Astaxanthin four years ago when I started racing triathlons, which involves two things: heavy exercise and long-term sun exposure. I was looking for a product that would help my performance in these two areas. My solution was Natural Hawaiian Astaxanthin. Once I started using it, I noticed a significant improvement in overuse injuries as well as long-term sun exposure. Antioxidants are the secret to training, performance, and recovery, and Natural Hawaiian Astaxanthin is packed full of high-quality antioxidants." *(Tim Marr, Honolulu, Hawaii, winner of the Pan American Long Distance Triathlon, which included a new swim course record)*

- "My sport of free-diving is physically demanding and I train in very extreme conditions. I religiously started taking Natural Hawaiian Astaxanthin in 2003 on the premise that it would take care of the free radicals in my system that bind to the oxygen molecules that would hinder oxygen uptake. Once I started taking Hawaiian Astaxanthin I noticed an overall change in my health. I was getting sick far less than in previous years. Colds and flus, which are potential problems during training and competition as I push my body to its limits, had become a non-issue for me. I was noticing changes in my training itself. During dives I found that on my ascent I was getting far less fatigued and there was absolutely no lactic acid build-up in my quadriceps. I was recovering from the dive much quicker on the surface, which means I am able to catch my breath far quicker than before. I conducted my own personal experiment and stopped using Hawaiian Astaxanthin to see what effects this would have on my body and training. During dives to 200 feet I was experiencing much fatigue in my quadriceps on my ascent. I was also taking much longer to recover from the dives on the surface than while I was taking Astaxanthin. This reinforced my belief that Hawaiian Astaxanthin really was making a difference in my dives and performances." ***(Deron Verbeck,*** *America's Top-Ranked Free-Diver, Kailua-Kona, Hawaii)*

## ACCLAIM FROM JOINT-ACHERS

- I was a collegiate athlete and have had a lot of problems with joint pain in my hands for years. In fact, it gets so bad that I'm unable to hold a newspaper for longer than five minutes without my hands and fingers getting sore. I started taking Hawaiian Astaxanthin about five years ago and since taking it my hands and fingers are 90% better. I started seeing results after the first two months. I have tried several different competitive products, and within two weeks my hands were just as bad as prior to taking Hawaiian Astaxanthin. I will definitely never use anything else! I'm glad I finally found something that works. ***(Mark Vieceli,*** *Director, Sales, Marketing and Business Development, Capsugel, Greenwood, South Carolina)*

- I was a competitive swimmer from the age of 3 through 18, then continued on a limited basis until age 24. After all these years of very heavy training, which included 4 hours daily of pool workouts and weight training, working as a lifeguard, competing in rough water swims, training in Junior Lifeguards, and surfing for fun – I developed severe tendonitis in both shoulders and both knees.

My tendonitis started flaring up when I was 14 years old; it got so bad within that same year that I dropped from being number 10 in the nation in sprint events to not even being on the top 50 list. I slept with ice packs on my shoulders and missed a lot of swim workouts. It was difficult to stand up from a crouching position, as the tendonitis in my knees was very painful; walking was painful, too. Tendonitis is the reason why I passed up scholarships for college and ultimately stopped competing.

I began taking Hawaiian Astaxanthin at age 29. I started with one, then two capsules daily. It took about four months for my tendonitis to heal to the point that I did not have pain or notice that it was ever there. Now it is two and a half years later. I'm still taking two capsules daily, and still no pain in my shoulders or knees. I have not altered any of my daily routines, diet, or exercise.

I directly attribute my use of Hawaiian Astaxanthin to the healing of my tendonitis. I had this condition for fifteen years, and nothing I did, didn't do, or tried ever worked. I wish this product was around when I was 14 years old, but I'm happy that I have it now. *(**Nicholle Davis**, Kailua-Kona, Hawaii)*

## SOOTHING STORIES FROM SUNBURNERS

- Robert Childs, M.D. who was born in Hawaii had always been extremely sun sensitive and would begin to burn in as little as half an hour of sun exposure. He reported that after taking Natural Hawaiian Astaxanthin he could go out in the midday sun for four hours without burning. He found that Astaxanthin "changed his life" by giving him not only more sunburn-resistant skin but also less painful joint soreness and stiffness.

# REFERENCES

Aoi, W., et al. (2003). Astaxanthin limits exercise-induced skeletal and cardiac muscle damage in mice. *Antioxidants & Redox Signaling*. 5(1): 139–44.

Awamoto, T., et al. (2000). Inhibition of low-density lipoprotein oxidation by Astaxanthin. *Journal of Atherosclerosis Thrombosis*. 7(4): 216–22.

Bagchi, D. (2001). Oxygen free radical scavenging abilities of vitamins C, B, B-carotene, pycnogenol, grape seed proanthocyanidin extract, Astaxanthin and BioAstin in vitro. Access at www.astaxanthin.org

Beutner, S., et al. (2001). Quantitative assessment of antioxidant properties of natural colorants and phytochemicals: Carotenoids, flavonoids, phenols and indigoids. The role of beta-carotene in antioxidant functions. *Journal of the Science of Food and Agriculture*. 81: 559–68.

Chan, K.C., et al. (2009). Antioxidative and anti-inflammatory neuroprotective effects of Astaxanthin and canthaxanthin in nerve growth factor differentiated PC12 cells. *Journal of Food Science*. 74(7): H225–31.

Chang, C.H., et al. (2010). Astaxanthin secured apoptotic death of PC12 cells induced by beta-amyloid peptide 25-35: Its molecular action targets. *Journal of Medicinal Food*. 13(3): 548–56.

Curec, GD., et al. (2010). Effect of Astaxanthin on hepatocellular injury following ischemia/reperfusion. *Toxicology*. 267(1–3): 147–53.

Fassett, R.G., et al. (2008). Astaxanthin versus placebo on arterial stiffness, oxidative stress, and inflammation in renal transplant patients: A randomized, controlled trial. *BMC Nephrology*. 9: 17.

Grangaud, R. (1951). "Research on Astaxanthin, a New Vitamin A Factor." Doctoral Thesis at University of Lyon, France. Available on the US Patent and Trademark Office website at www.uspto.gov

Gross, G.J., et al. (2005). Acute and chronic administration of disodium disuccinate Astaxanthin produces marked cardioprotection in dog hearts. *Molecular and Cellular Biochemistry*. 272: 221–27.

Hussein G., et al. (2006). Anti-hypertensive potential and mechanism of action of Astaxanthin II. Vascular reactivity and hemorheology in spontaneously hypertensive rats. *Biological and Pharmaceutical Bulletin*. 28(6): 967–71.

Hussein, G., et al. (2005). Anti-hypertensive and neuroprotective effects of Astaxanthin in experimental animals. *Biological and Pharmaceutical Bulletin*. 28(1): 47–52.

Iwamoto, T., et al. (2000). Inhibition of low-density lipoprotein oxidation by Astaxanthin. *Journal of Atherosclerosis Thrombosis*. 7(4): 216–22.

Izumi-Nagai, K., et al. (2008). Inhibition of choroidal neovascularization with an anti-inflammatory carotenoid Astaxanthin. *Investigative Opthamology & Visual Science*. 49(4): 1679–85.

Kim, J.H., et al. (2010). Astaxanthin improves stem cell potency via an increase in the proliferation of neural progenitor cells. *International Journal of Molecular Sciences*. 11(12): 5109–19.

Kim, J.H., et al. (2009). Astaxanthin inhibits H202-mediated apoptotic cell death in mouse meural progenitor cells via modulation of P38 and MEK signaling pathways. *Journal of Microbiology and Biotechnology*. 19(11): 1355–63.

Lauver, D.A., et al. (2005). Disodium disuccinate Astaxanthin attenuates complement activation and reduces myocardial injury following ischemia/reperfusion. *Journal of Pharmacology and Experimental Therapeutics*. 314: 686–92.

Lee, D.H., et al. (2011). Astaxanthin protects against MPTP/MPP+-induced mitochondrial dysfunction and ROS production in vivo and in vitro. *Food and Chemical Toxicology.* 49(1): 271–80.

Liu, X., et al. (2009). Astaxanthin inhibits reactive oxygen species-mediated cellular toxicity in dopaminergic SH-SY5Y cells via mitochondria-targeted protective mechanism. *Brain Research.* 1254: 18–27.

Lorenz, T. (2002). Clinical trial indicates sun protection from BioAstin supplement. Accessed at www.astaxanthin.org

Lu, Y.P., et al. (2010). Neuroprotective effect of Astaxanthin on H(2)O(2)-induced neurotoxicity in vitro and on focal cerebral ischemia in vivo. *Brain Research.* 1360: 40–8.

Mason, P., et al. (2006). Refecoxib increases susceptibility of human LDL and membrane lipids to oxidative damage: A mechanism of cardiotoxicity. *Journal of Cardiovascular Pharmacology.* 47(1): S7–S14.

Massonet, R. (1958). "Research on Astaxanthin's Biochemistry." Doctoral Thesis at University of Lyon, France. Available on the US Patent and Trademark Office website at www.uspto.gov

Miyawaki, H., et al. (2008). Effects of Astaxanthin on human blood rheology. *Journal of Clinical Biochemistry Nutrition.* 43(2): 9–74.

Miyawaki, H., et al. (2005). Effects of Astaxanthin on human blood rheology. *Journal of Clinical Therapeutics & Medicines.* 21(4): 421–29.

Monrov-Ruiz, J., et al. (2011). Astaxanthin-enriched diet-reduces blood pressure and improves cardiovascular parameters in spontaneously hypertensive rats. *Pharmacological Research.* 63(1): 44–50.

Nakagawa, K., et al. (2011). Anti-oxidant effect of Astaxanthin on phospholipid peroxidation in human erythrocytes. *The British Journal of Nutrition.* 105(11): 1563–71.

Nakao, R., et al. (2010). Effect of Astaxanthin supplementation on inflammation and cardiac function in BALB/mice. *Anticancer Research.* 30: 2721–25.

Nishioka, Y., et al. (2011). The anti-anxiety-like effect of Astaxanthin extracted from Paracoccus carotinifaciens. *Biofactors.* 37(1): 25-30.

Palozza, P., et al. (2009). Growth-inhibiting effects of the Astaxanthin-rich Haematococcus pluvialis in human colon cancer cells. *Cancer Letters.* 283(1): 108–17

Park, J.S., et al. (2010). Astaxanthin decreased oxidative stress and inflammation and enhanced immune response in humans. *Nutrition and Metabolism.* 7: 18.

Pashkow, F.J., et al. (2008). Astaxanthin: A novel potential treatment for oxidative stress and inflammation and cardiovascular disease. *The American Journal of Cardiology.* 101(10a): 58d–68d.

Satoh, A., et al. (2009). Preliminary clinical evaluation of toxicity and efficacy of a new Astaxanthin-rich Haematococcus pluvialis extract. *Journal of Clinical Biochemistry and Nutrition.* 44(3): 280–84.

Sears, W., (2010). *Prime-Time Health.* A proven plan to live happier, healthier, and longer. New York: Little, Brown.

Sears, W., and Sears, J. (2012). *The Omega-3 Effect.* Everything you need to know about the supernutrient for living longer, happier, and healthier. New York: Little, Brown.

Shen, H., et al. (2009). Astaxanthin reduces ischemic brain injury in adult rats. *The FASEB Journal.* 23(6): 1958–68.

Shimidzu, N., et. al. (1996). Carotenoids as singlet oxygen quenchers in marine organisms.

*Fisheries Science*. 62(1): 134–37.
Shiratori, K., et al. (2005). The effects of Astaxanthin on accommodation and asthenopia – efficacy identification study in healthy volunteers. *Clinical Medicine*. 21(6)J: 637–50.
Spiller, G., et al. (2006). Effect of daily use of natural Astaxanthin on C-reactive protein. *Health Research and Studies Center*, Los Altos, CA. Access at www.astaxanthin.org
Sun, Z., et al. (2011). Protective actions of microalgae against endogenous and exogenous AGEs in human retinal pigment epithelial cells. *Food and Function*. 2(5): 251–8.
Trimeks Company Study (2003). Access at www.astaxanthin.org
Wang, H.Q., et al. (2010). Astaxanthin upregulates heme oxygenase-1 expression through ERK 1/2 pathway and its protective effect against beta-amyloid-induced cytotoxicity in SH-SY5Y cells. *Brain Research*. 1360: 159–67.
Yamashita, E., et al. (2006). The effects of a dietary supplement containing Astaxanthin on skin condition. *Carotenoid Science*. 10: 91–95.
Yazaki, K., et al. (2011). Supplemental cellular protection by a carotenoid extends lifespan via INS/IGF-1 signaling in C. elegans. *Oxidative Medicine and Cellular Longevity*.
Yoshida H., et al. (2010). Administration of Natural Astaxanthin increases serum HDL-cholesterol and adiponectin in subjects with mild hyperlipidemia. *Atherosclerosis*. 209: 520–3.

## NOTES

## ABOUT THE AUTHOR

**William Sears, MD** is the author of over 40 books and is widely recognized as America's leading pediatrician. A book series written with his wife Martha Sears, R.N. called the "Sears Parenting Library" featuring the "Attachment Parenting Philosophy" has become the parenting bible to a new generation of young parents. Following his passion for family health and longevity, in this book Dr. Sears reveals one of the top nutrition secrets of the sea: Astaxanthin. Dr. Sears has been on over 100 television shows including *Oprah Winfrey, Dr. Phil, 20/20, Good Morning America, CBS This Morning, CNN, The Today Show* and *The Doctors*. Dr. Sears is a fellow of both the American Academy of Pediatrics and the Royal College of Pediatricians. He still finds the time to actively practice medicine in San Clemente, California, where he lives in a home frequently visited by his eight children and a growing number of grandchildren.